AF739883

Safety Mindset

Workplace safety is a simple construct geared towards one philosophy: every person deserves to return home safely to their loved ones. Drawing from experience in education and retail industries, this book illustrates how quickly ordinary decisions can dramatically alter our future paths. Through compelling real-world examples, including a powerful light switch demonstration that transformed a child's bedroom into a landscape of potential hazards, the author reveals the fragile nature of safety and how easily it can be compromised by thoughtless actions.

This book offers a strategic framework and practical resources needed to transform safety culture through the power of better decisions. Readers will discover the "Safe 2 Home" approach that transforms safety management from a rigid science into an art of decision-making influence. The book presents over 120 ready-to-implement toolbox talks covering critical safety topics from fall arrest systems to equipment operation, each designed to create meaningful emotional connections with safety protocols. Safety leaders can develop and possess practical tools to build a proactive safety culture where thoughtful decisions actively save and improve lives every day.

Safety Mindset: Conversations That Work will appeal to professionals in the field of occupational health and safety, including safety managers, supervisors, HR personnel and anyone responsible for workplace well-being.

Safety Mindset

Conversations That Work

Justin Hinks

CRC Press
Taylor & Francis Group
Boca Raton London New York

CRC Press is an imprint of the
Taylor & Francis Group, an **informa** business

Front cover image: andresr/iStock

First edition published 2026
by CRC Press
2385 NW Executive Center Drive, Suite 320, Boca Raton FL 33431

and by CRC Press
4 Park Square, Milton Park, Abingdon, Oxon, OX14 4RN

CRC Press is an imprint of Taylor & Francis Group, LLC

ISBN: 978-1-041-16819-5 (hbk)
ISBN: 978-1-041-15720-5 (pbk)
ISBN: 978-1-003-68651-4 (ebk)

DOI: 10.1201/9781003686514

Typeset in Times
by Newgen Publishing UK

Contents

Foreword

In our workplaces, safety is often an unsung hero—a foundational pillar that ensures everyone makes it back home to their loved ones. Yet, its fragility often goes unnoticed until it is tested. This book, *Safety Mindset*, is a heartfelt and practical guide from an experienced Safety Director who has not only walked the floors of warehouses and retail spaces but has also explored the deeper implications of decision-making for safety.

Through personal anecdotes, innovative insights, and a deep commitment to the vision of "Safe 2 Home,“ the author challenges us to rethink how small, everyday choices impact the lives of individuals and organizations. From flipping a light switch to calling attention to overlooked hazards, these seemingly minor acts remind us that safety is not merely a policy—it's a shared responsibility, a culture, and an art.

More than a collection of Toolbox Talks, this book is a call to action for leaders to be visible, inspiring, and dedicated to fostering connections and challenging conventions. It's about teaching others the power of decisions and making safety a centerpiece of organizational success.

As you delve into the pages of *Safety Mindset: Conversations That Work*, prepare to be inspired and equipped with the tools to create a safer world—not just for yourself but for every individual you lead. Let the lessons within remind you of the ultimate goal: making it Safe 2 Home.

Sincerely,
Nick Passolano—
Sr. Client Service Manager,
CWG Insurance

Preface

Safety Mindset was born from a sincere desire to truly connect with the hearts of the individuals I was entrusted to protect. When I was first thrust into a safety role, I quickly found myself out of airspeed and on a glide slope to nowhere. Teaching adults about safety, I realized, would take far more than instruction, lesson plans, or group activities. It would take connection.

That pursuit led me to a driver training session—one that changed everything. We were discussing the importance of making it home safely. I asked the group to consider who and what would be impacted if they didn't make it home today. One gentleman lit up the room as he spoke about his grandkids. That was it. That was the connection I had been searching for. Safe 2 Home was born.

When we discover someone's spark—their reason for being—it becomes incredibly easy to connect safety messages to them in a lasting, meaningful way.

I've had the privilege of working with young people, incredible teachers, inspiring leaders, and dedicated parents. I've also worked alongside my family—on the farm and in the office. Connecting with people has always been a strength of mine, but this was different. This idea had the potential to change how people connect with others. It had to be developed further.

Safety Mindset is designed to be an essential tool for safety leaders. Its purpose is simple: to help frontline safety professionals deliver messages that resonate. It offers not only the power to connect but also a repeatable method for doing so. Every safety manager struggles with their first safety talk—so that's where we begin.

This book introduces and explains the Safe 2 Home process and provides over 120 discussion prompts to support new and seasoned safety leaders alike.

This work would not have been possible without the people who shaped it:

- **Kyle**—Thank you for making safety real. You showed me how to live and breathe it.
- **Ari**—Your knowledge and dedication to helping others grow is unmatched in our industry.
- **Robyn**—Your edits, suggestions, and unwavering support have been invaluable.
- **Nick**—Thank you for believing. We did it, my friend.

I sincerely hope you enjoy *Safety Mindset: Conversations That Work.*[1] I believe it will make a real difference—for those new to safety and for seasoned veterans looking to be re-energized. As you read, I encourage you to reflect on the teachers, mentors, and professionals who helped you connect to your work and make a difference in the lives of others.

NOTE

1 Professionally Microsoft Co-Pilot was used to assist in the structure of the individual discussion Prompts (Toolbox Talks). It was also used to review language, grammar, and punctuation. Any language modification was reviewed and rephrased to align with the works overarching themes. Specific sources are cited where appropriate.

About the Author

Justin Hinks, from Tea, South Dakota, USA, holds masters degrees in educational administration and business administration with a focus on Project Management. With over a decade of experience in safety management, he has spent the last eight years as the Director of Safety for Furniture Mart USA, overseeing 70 locations and nearly 1,300 employees across six states. Certified as a Safety Manager through NASP, in Incident Command through FEMA, and trained as a DOT inspector (MIPS), Justin's professional philosophy centers on proactive prevention and continuous improvement. His focus is on creating awareness and improving communication around safety, ensuring that every family member makes it home safely today, tomorrow, and into the future. He believes that coaching and mentoring are essential to eliminating workplace safety incidents. Recognized for his contributions to workplace safety, Justin is a member of several professional organizations and has spoken at several industry conferences. He is dedicated to safety management innovation and helping organizations achieve their safety goals. His "Safe 2 Home" mission is driven by his commitment to his wife Angie, his son Wes, and their two dogs, Gemma and Sylvee. In his free time, he enjoys working on the family farm ground, restoring old trucks, and spending quality time with his family.

1 The Decision for Safety

After years of working in education and the fast-paced retail industry, I've realized that we, as co-workers, share a common mission: to make it home safely to our family, friends, and pets. Each of us has a unique vision or dream of what life should be, and that idea is more fragile than we might want to admit. As a Safety Director, I've seen firsthand how quickly the simplest decision can alter an individual's or group's vision for their future.

Safety is incredibly difficult to maintain and frighteningly simple to destroy. Safety Managers worldwide will never run out of opportunities to improve the safety of others. In the end, making people safe isn't a science; it's the art of helping others make better decisions. Decisions create safer processes, facilities, and workers.

Reflecting on safety, I've learned a valuable lesson from my 13-year-old son. He taught me the importance of teaching others about the power of decision-making. One day, after he made a mistake, I asked him why he made that decision. His answer was simple: "I didn't think about it." To illustrate the importance of thinking through actions, I flipped the light switch in his room, making it nearly pitch black. This small decision changed our perspective on everything in the room, turning toys into trip hazards and a glass of water into danger to the iPad on the desk.

The light switch conversation led to the birth of the phrase "Safe 2 Home." What if we could teach others the power of decisions? What if we could show them the importance of the small actions, we take every day?

This has become my vision, my ideal, my dream: to help my family, friends, and colleagues understand the power of decisions and their impact on safety. All industries, especially retail and warehousing, are ready for change. The Safety Mindset aims to help safety managers influence cultural shifts with safety as the main driver. The *Safety Mindset: Conversations That Work* is the first in a series of writings to encourage others to spend a little more time and effort thinking about making it Safe 2 Home.

BE VISIBLE

Safety leadership starts with you. Your actions, words, and inaction will be judged. Each decision you make impacts those around you. Be visible and make safe decisions.

DOI: 10.1201/9781003686514-1

Persistently and diligently get to know those in your charge. By getting to know them personally, we create opportunities to connect to their Safe 2 Home mission.

Here's an example of being visible. Walking through our warehouse one day, I encountered a young man who wasn't following a safety policy with a piece of machinery. The likelihood of injury was low, but I decided to be "visible." We talked briefly about his workload and weekend plans. He and his girlfriend were going to an event. I asked him how she would feel about him ignoring our safety procedures. He became quiet and said he would remedy the situation. I reminded him of Safe 2 Home and suggested we call his girlfriend. When she answered, I mentioned the safety risk and its potential impact on their weekend. This conversation made the safety risk more serious for him.

Can you commit to being visible? Can you walk the floor, field, or yard and coach by catching? The following collection of more than 120 Toolbox Talks (TBTs) will assist any safety leader with at least one training session a week for two years. Use this collection as a starting point to launch your own safety decision culture.

INSPIRE

Leadership isn't about giving orders or providing instruction; it's about influence. Once you become a visible symbol of caring, concern, and compassion, you create opportunities to connect. Connections allow you to inspire. Without the ability to connect personally with what is important to others, we cannot show them what is important to us. If we cannot show them what is important, we cannot expect to shift their mindset.

Here's an example anyone can use today. Compliment the next person you see at a gas station on their pickup truck. This simple act creates a connection. A smile and emotional gratefulness enter the individual when they are complimented.

Once we create the connection, we begin to have influence. John Maxwell talks about this in his books and seminars. Watching his seminars, John creates a natural connection with everyone he meets by shaking their hand, inviting them into his space, and asking questions (Maxwell, 2010). This disarms them and makes them feel important, increasing the value of his opinion.

As safety leaders, we can learn from Mr. Maxwell. How do we inspire others to think about safety the way we do? Once the connection is established, we can use our visibility and proximity to demonstrate how much we value the connection. This encourages change and inspires a commitment to safety.

CHALLENGE EVERYTHING

Organizational leaders need constant communication from the front line. Production and quality are easy to measure, but how do we measure the impact of safety on organizational performance? Many organizations look to their DART or MOD ratings, but these numbers tell only a small fraction of the story.

Hundreds of decisions are made daily that could lead to exponential growth or sudden collapse. Leaders must allow staff to challenge everything. This means constantly re-evaluating processes, productions, technology, and facilities. It requires letting go of the "how we have always done it" mentality and valuing others more than ourselves.

An example of challenging everything comes from the escape velocity of a golf ball. A former colleague realized that surface tension was the secret to creating a high traction, slip-resistant surface. By applying the principles of a dimpled golf ball, they created a coating that became high traction when wet and smooth when dry. They now are a partner in what I would call a Revolutionary Flooring Company.

Another example comes from manufacturing and shipping. During the COVID pandemic, the availability of goods slowed drastically, leading to inflation. Some manufacturers challenged the high cost of transportation by condensing packaging, which revolutionized their entire companies.

Leaders must value the opinion and talents of their staff. It makes sense to invest in their ability.

DEDICATION

Safety leadership is about understanding the value of coaching decisions. The best safety leaders follow the mindset of being visible, inspiring, and challenging everything. It is this dedication that sets safety leaders apart. Dedication to teaching "decision-making" makes the mission of making it Safe 2 Home possible.

As you continue to grow as a safety leader and professional, write down your own personal Safe 2 Home mission? Who's waiting for you? Who's counting on you? Who will be here tomorrow if you don't show up? We all have our own reasons; I challenge you to live yours.

For me, I dedicate myself to making it home safely to my wonderful wife, my smart and witty son, and my two dogs, Gemma and Sylvie, who keep our home safe each and every day. I would also like to thank my parents, in laws, and extended family for believing in me and this work. It has allowed me to share in the lives of thousands of young people, developing professionals, and leaders within our communities. This book would not be possible without you.

THE TOOLBOX TALK

The TBT originated in the trade industry, where workers would gather to discuss daily activities. In order to maximize the value of this time, a full day's tasks were reviewed. What teams would be tasked with doing each job, and how those jobs and teams may cross paths throughout the day. Each task would be considered a "job," and each job would have an unique JHA or Job Hazard Assessment (Occupational Safety and Health Administration, OSHA Guidelines for Job Hazard Analysis, 2025). This is where the TBT really starts to stand apart from project management.

Each job and employees unique JHA would identify hazard exposure and precautionary practices to mitigate the risk. Personal Protective Equipment or PPE would be identified and communication requirements established. The daily TBT would ensure the team understands and clarifies any overlapping concerns to other teams. This informal communication is intentionally done with all staff, to allow for anyone to recognize and point out potential dangers.

When designing the TBT it's important to "connect" to the audience and make the training relevant and impactful. There are several basic design techniques that will establish a framework for TBTs that work. The primary framework should identify the topic and risks and ensure, through clarification and engagement, that the content is understood. Failing to confirm the message was received, nullifies the potential benefit of our work as safety professionals.

If the primary focus is to identify risks, we should also be sure to note, how we intended to engage the group in the conversation. Without engagement, we fail to "connect" content to the audience. After creating a means for engagement, we need to ensure content conveyed and discussed is retained. An activity or test is a great way to prove understanding. Said differently, there are three key components: sharing of knowledge, fostered engagement, and a knowledge check. By including these three items in your toolbox framework you will ensure knowledge conveyed is knowledge retained.

This collection of TBTs is designed to inspire and challenge you to be a safety leader, be visible, foster connections, inspire, and encourage others to challenge everything. The series starts off with a basic framework for TBT creation—"The Art of the TBT." Many other trainings follow and are organized alphabetically. They cover a variety of basic topics for general industry. As a certified safety Manager, the content of these TBT's is built upon my interpretation of regulations as applied to industries, which I am experienced in. When training your team, be sure to consult your OSHA guidelines, city, state, and federal regulations before communicating specific safety instruction.

Feel free to use and distribute these TBTs within your organization. I have included 124 TBTs to set the stage for you. A TBT a week for more than two years. Use these as a foundation in your safety program to help others refocus on what is most important: making it Safe 2 Home.

2 Conversations That Work

ART OF THE TBT

Toolbox talks are short, informal safety meetings that focus on specific safety issues related to the job at hand. They are essential for reinforcing safety practices, raising awareness about potential hazards, and fostering a culture of safety in the workplace.

Key Components of a Toolbox Talk:

1. **Topic Selection**: Choose a relevant and timely topic based on current site conditions, recent incidents, or upcoming tasks. Common topics include PPE usage, fall protection, and hazard communication.
2. **Preparation**: Gather necessary materials and information, such as safety data sheets (SDS), equipment manuals, or visual aids. Ensure you understand the topic thoroughly.
3. **Introduction**: Start with a brief introduction that explains the purpose of the talk and why the topic is important.
4. **Key Points**: Outline the main points you want to cover. Use clear, straightforward language and avoid technical jargon.
5. **Engagement**: Encourage participation by asking open-ended questions and inviting workers to share their experiences and concerns.
6. **Visual Aids**: Use posters, handouts, videos, or demonstrations to enhance understanding and retention.
7. **Summary**: Recap the key points and ensure everyone understands the main messages.
8. **Documentation**: Record attendance and the topics discussed for future reference and compliance purposes

Creating and Fostering Discussion:

1. **Scenario-Based Questions**: Present hypothetical scenarios related to the topic and ask employees how they would respond. This encourages critical thinking and practical application of safety protocols.
2. **Personal Experiences**: Invite employees to share their own experiences related to the topic. This can highlight real-world challenges and solutions.

DOI: 10.1201/9781003686514-2

3. **Interactive Activities**: Organize small group activities where employees can discuss and solve safety-related problems together.
4. **Open Forum**: Allow time for an open forum where employees can ask questions and discuss concerns about the topic.
5. **Guest Speakers**: Invite experts to talk about specific safety issues and their experiences.

Knowledge Check:

1. **Quizzes**: Conduct short quizzes at the end of the talk to reinforce key points and assess understanding.
2. **Practical Demonstrations**: Have employees demonstrate the correct use of PPE or other safety procedures.
3. **Feedback Loop**: Encourage employees to provide feedback on the talk and suggest topics for future discussions.
4. **Follow-Up**: Review the key points in subsequent meetings to reinforce the information and address any lingering questions or concerns.

By following these guidelines, you can create effective toolbox talks that engage your team, reinforce safety practices, and foster a culture of safety in the workplace. Remember, the goal is to make safety a routine part of your daily operations. Stay safe and keep the conversation going!

AED "DEFIBRILLATOR" SAFETY

Good morning, team. Today, I want to discuss a critical topic that can save lives in emergencies—defibrillators and Automated External Defibrillators (AEDs). Knowing how to use these devices and understanding their importance can make a significant difference in a cardiac emergency (AED implementation Guide, 2020).

What Is a Defibrillator?
A defibrillator is a device that delivers an electric shock to the heart to restore a normal heartbeat. It is used in cases of sudden cardiac arrest (SCA), where the heart stops beating effectively.

What Is an AED?
An Automated External Defibrillator (AED) is a portable device that checks the heart's rhythm and can send an electric shock to the heart to try to restore a normal rhythm. AEDs are designed to be used by laypersons with little or no training.

Importance of AEDs:

1. **Life-Saving Potential:** AEDs can significantly increase the chances of survival for someone experiencing sudden cardiac arrest. Immediate use of an AED can double or triple survival rates.

2. **Ease of Use:** AEDs are user-friendly and provide clear, step-by-step voice instructions. They are designed to be used by anyone, even without medical training.
3. **Accessibility:** AEDs are commonly found in public places such as airports, schools, and workplaces. Knowing their locations can be crucial in an emergency.

How to Use an AED:

1. **Call for Help:** Immediately call emergency services if someone collapses and is unresponsive.
2. **Turn on the AED:** Open the AED and turn it on. Follow the voice prompts.
3. **Attach the Pads:** Place the adhesive pads on the person's bare chest as shown in the diagrams on the pads.
4. **Analyze the Heart Rhythm:** The AED will automatically analyze the heart rhythm. Ensure no one is touching the person during this time.
5. **Deliver the Shock:** If the AED advises a shock, ensure everyone is clear of the person and press the shock button.
6. **Continue CPR:** After the shock, continue CPR until emergency responders arrive or the person shows signs of life.

Thank you for your attention. Remember, being prepared and knowing how to use an AED can save lives. Let's stay vigilant and ready to act in case of an emergency.

AIR TOOL SAFETY

Pneumatic and air-powered tools are efficient and powerful, but they can be hazardous if not used properly. This talk will cover essential safety precautions for using these tools, including hoses, tools, and blow guns (Hand and Portable Powered Tools 29 CFR 1910.242, 2025).

Safety Precautions:

1. **Inspect Before Use**
   - **Hoses**: Check for cracks, leaks, and wear. Ensure connections are secure.
   - **Tools**: Inspect damage, loose parts, and proper operation.
   - **Blow Guns**: Ensure the nozzle is intact and the trigger mechanism works correctly.
2. **Proper Hose Management**
   - **Avoid Kinks and Bends**: Prevent kinks and sharp bends in hoses to maintain airflow and prevent damage.
   - **Secure Hoses**: Keep hoses away from walkways to prevent tripping hazards. Use hose reels or hangers for storage.
   - **Pressure Ratings**: Use hoses rated for the specific pressure of the tool being used.

3. **Safe Tool Operation**
   - **Follow Manufacturer's Instructions**: Always use tools according to the manufacturer's guidelines.
   - **Use Appropriate PPE**: Wear safety glasses, hearing protection, and gloves.
   - **Disconnect When Not in Use**: Disconnect tools from the air supply when not in use or when changing accessories.
4. **Blow Gun Safety**
   - **Use Proper Nozzles**: Use blow guns with safety nozzles that limit pressure to safe levels.
   - **Avoid Directing Air at People**: Never point a gun at yourself or others. High-pressure air can cause serious injuries.
   - **Control Air Pressure**: Use a pressure regulator to control the air pressure to safe levels.
5. **Preventing Accidental Discharge**
   - **Secure Connections**: Ensure all connections are tight and secure to prevent hoses from detaching.
   - **Check for Leaks**: Regularly check for air leaks and repair them immediately.
   - **Use Safety Clips**: Use safety clips or retainers on hose connections to prevent accidental disconnection.
6. **Maintenance and Storage**
   - **Regular Maintenance**: Perform regular maintenance on tools and hoses as recommended by the manufacturer.
   - **Proper Storage**: Store tools and hoses in a clean, dry place to prevent damage and deterioration.
   - **Drain Moisture**: Regularly drain moisture from air compressors and hoses to prevent rust and damage.

Discussion Points:

1. What are some common issues you have encountered with pneumatic tools and hoses?
2. How do you ensure that hoses are properly managed and stored in your work area?
3. Can you describe the steps you take to inspect tools and hoses before use?
4. What PPE do you use when operating pneumatic tools, and why is it important?
5. How do you handle and store blow guns to ensure safety?
6. What procedures do you follow to prevent accidental discharge of pneumatic tools?

Conclusion

By following these safety precautions and discussing potential hazards, we can minimize the risks associated with pneumatic and air-powered tools and ensure a safer working environment. Always stay vigilant and report any safety concerns to your supervisor.

Your Safe 2 Home mission starts and ends with YOU.

AIR QUALITY

1. **Understanding Air Quality**:
    - Good air quality is essential for a healthy and productive work environment. Poor air quality can lead to health issues such as headaches, fatigue, and respiratory problems.
2. **Prevention Measures**:
    - Ensure proper ventilation in work areas by regularly maintaining HVAC systems and using air purifiers if necessary. Avoid using products that emit strong odors or pollutants.
3. **Reporting Concerns**:
    - If you notice any unusual odors, visible dust, or experience symptoms like headaches or dizziness, report these concerns immediately to your supervisor or the designated safety officer.
4. **Personal Responsibility**:
    - Keep your workspace clean and free of clutter. Dispose of waste properly and avoid blocking air vents. Use eco-friendly and low-emission products whenever possible.

Discussion Points:

- **How can we improve ventilation in our work areas?**
- **What steps can we take to ensure everyone is aware of the importance of reporting air quality concerns?**

The most common concern is Particulate Matter:

- Dust, dirt, and other fine particles can be inhaled, causing respiratory problems and exacerbating conditions like asthma.

Report on all Air Quality concerns.

ALLERGIES AND EXPOSURES

1. **Common Allergens**:
    - Common workplace allergens include pollen, dust mites, mold, chemicals, and certain foods. These can cause allergic reactions that may affect your health and productivity.
2. **Symptoms of Allergic Reactions**:
    - Symptoms can vary but often include sneezing, coughing, itchy or watery eyes, skin rashes, and difficulty breathing. Severe reactions, such as anaphylaxis, require immediate medical attention.
3. **Prevention Measures**:
    - Keep work areas clean and well-ventilated. Use air purifiers if necessary and avoid using products that emit strong odors or chemicals. Be aware of any known allergens and take steps to minimize exposure.

Reporting Concerns:

- If you experience symptoms or notice potential allergens in the workplace, report your concerns immediately to your supervisor or the designated safety officer. Prompt reporting can help address issues before they become serious.

First Aid Measures:

- **Mild Reactions**:
  - For mild symptoms like sneezing or itching, over-the-counter antihistamines can be helpful. Ensure the affected person is comfortable and monitor their condition.
- **Severe Reactions (Anaphylaxis)**:
  - Call 911 immediately if someone is experiencing severe symptoms such as difficulty breathing, swelling of the face or throat, or a rapid pulse. If the person has an epinephrine auto-injector (EpiPen), help them use it. Keep the person calm, have them lie down, and elevate their legs. Do not give them anything to drink.

Discussion Points:

- **What are some common allergens you have encountered at work?**
- **How can we improve our current practices to reduce exposure to allergens?**

AMPUTATIONS

Hello everyone! Today, we are going to discuss a very serious topic: workplace amputations. These injuries can have life-altering consequences, so it is crucial to understand the common causes, the importance of proper training, and the role of machine guarding in preventing such incidents (General Safety Requirements for All Machines 29.CFR 1910.212, 2025).

Key Points:

1. **Common Causes of Amputations**:
   - **Machine Accidents**: Amputations often occur due to entanglement or crushing incidents involving heavy machinery and equipment. Workers can get caught in moving parts, belts, gears, or other pinch points.
   - **Lack of Safeguards**: The absence or improper maintenance of safety guards on machines can expose workers to hazardous areas, increasing the risk of amputations.

- **Inadequate Training**: Insufficient training on machine operations and safety protocols can lead to improper use of equipment, resulting in severe injuries.

2. **Importance of Training**:
   - Proper training is essential to ensure that employees understand how to safely operate machinery and recognize potential hazards. Training helps prevent accidents by educating workers on the correct procedures and safety measures.
   - Regular refresher courses and hands-on training sessions can reinforce safety protocols and keep safety top of mind for all employees.
3. **Machine Guarding**:
   - Machine guards are critical for protecting workers from injuries caused by moving parts. Guards should be in place to prevent contact with dangerous areas, be securely attached, and not create new hazards.
   - Regular inspections and maintenance of machine guards are necessary to ensure they are effective and not compromised.

Discussion Points:

- **What are some examples of machine guarding in our workplace, and how do they help prevent injuries?**
- **How can we improve our training programs to ensure everyone understands the importance of machine safety?**

By staying informed and vigilant, we can significantly reduce the risk of amputations and create a safer work environment for everyone. Feel free to share your thoughts or any additional points you think should be included!

Together we can all make it **Safe 2 Home.**

ANGER: EMOTIONAL POISON

Anger, Frustration, Annoyance. These are powerful emotions we all experience, either at home or more specifically in our high-pressure work environments. It should be no surprise that these emotions do more than just affect your mood. They can directly impact your safety and the safety of others.

Places of work that experience higher levels of emotional tension, specifically anger and frustration, report significantly more workers' comp claims. But why? Because these negative emotions act like poison:

- They cloud our judgement.
- Slow our reaction time.
- Increase our acceptance of risk.
- Create a downward spiral of negativity with your colleagues.

The research also supports this: Dr. Daniel Vinson from the University of Missouri published similar findings. According to his report, nearly 1 in 10 injured people reported being irritable or angry prior to the incident. In other words, Anger can drastically increase your risk of injury (Vinson and Arelli, 2006). Similar findings can be seen in the automotive, construction, and sports environments.

Key Takeaways

Today we are going to focus on the cure for the poison. Emotional Control. Let's look at 4 strategies you can implement today to reduce your anger and frustration at work.

We have to be strong enough to:

1. "Name It." This means recognizing our emotional state, saying to ourselves where we are at then redirecting ourselves. This is like pulling the parking brake in our minds.
2. "Lock it in" is another easy drill. Start by mentally envisioning your standing in your favorite place, just you, and a metal locker with a lock. Now put the thing that's bugging you in the box and lock it up tight. Take a deep breath and mentally turn away from the concern and move on.
   This strategy works because it resets our mental space. Creating a sense of happiness and control over the situation. This reduces anxiety and calms the nervous system.
3. "Shoot Your Shot." This is a physical activity that helps burn off adrenaline and gives you back control. Think about that perfect basketball shooting position, set your feet, and practice your favorite jump shot. No ball needed. Focus on:
   - Your feet positioning
   - Is your elbow at 90%?
   - How does the ball roll off your fingers?

   By focusing in on this very specific physical movement and acknowledging a level of sensory feelings, we strategically push our brains out of the frustration loop.
4. "Speak up" sometimes we just need to say something to someone. Text or call a friend or family member. Ask them about their day. Give yourself permission to take an interest in them. If they ask how your day is. Share you're not in the best head space but are working to correct it.

This works, because emotionally we all want to be acknowledged and to belong. Most often our anger or frustration stems from a lack of feeling understood. Supercharging our emotional state with an empathetic conversation can help us bounce back.

Discuss

Which one of these seems like the best one to try? Does anyone do something like this already? Negative emotions like Anger are natural but must be managed. Let's commit to protecting ourselves and each other by recognizing the emotional poison we are feeding ourselves and each other and choosing a healthier recipe.

ARC FLASH SAFETY

Arc flashes are dangerous electrical events that can cause severe injuries or even fatalities. It's crucial to understand and follow safety practices to protect yourself and others when working around electrical equipment. This TBT is based on information found under NFPA 70E (NFPA 70E: Standard for Electrical Safety in the Workplace, 2021). Additional supporting information can be found in OSHA 29 CFR 1910.333 (29 CFR 1910.333, 2025).

1. **Understand the Hazards**
   - **What is an Arc Flash?** An arc flash is a sudden release of electrical energy through the air, resulting in an explosion.
   - **Potential Injuries**: Burns, hearing loss, blindness, and even death.
2. **Wear Appropriate PPE**
   - **Personal Protective Equipment**: Use flame-resistant clothing, insulated gloves, face shields, and hearing protection.
   - **Proper Fit**: Ensure PPE fits correctly and is in good condition.
3. **Follow Lockout/Tagout (LOTO) Procedures**
   - **De-energize Equipment**: Always de-energize electrical equipment before performing maintenance.
   - **LOTO**: Use LOTO procedures to ensure equipment remains de-energized.
4. **Maintain Safe Distances**
   - **Boundaries**: Establish and respect arc flash boundaries to keep a safe distance from live electrical parts.
   - **Warning Signs**: Use signs and barriers to alert others of arc flash hazards.
5. **Use Proper Tools and Equipment**
   - **Insulated Tools**: Use tools rated for electrical work to prevent accidental contact with live parts.
   - **Regular Inspections**: Inspect tools and equipment regularly for damage.
6. **Training and Awareness**
   - **Regular Training**: Ensure all personnel receive regular training on arc flash safety.
   - **Stay Informed**: Keep up to date with the latest safety standards and practices.

Conclusion

By following these safety practices, we can reduce the risks associated with arc flashes and ensure a safer working environment. Always stay vigilant and report any safety concerns to your supervisor.

ATTENDANCE AND PUNCTUALITY

Punctuality and attendance are crucial for maintaining a productive and efficient workplace. Being on time and present ensures that work flows smoothly and that everyone can rely on each other. This talk will cover the importance of punctuality and attendance and include open discussion points to ensure everyone understands their role in maintaining a reliable work environment.

Importance of Punctuality and Attendance
Being punctual and maintaining good attendance shows respect for your coworkers and the organization. It helps in meeting deadlines, maintaining productivity, and ensuring that everyone can depend on each other.

Discussion Point: How does punctuality and attendance impact our team's productivity and morale?

Benefits of Good Attendance:

1. **Reliability**: Consistent attendance builds trust and reliability among team members.
2. **Efficiency**: Being present and on time ensures that tasks are completed efficiently and on schedule.
3. **Professionalism**: Good attendance reflects a professional attitude and commitment to your job.
   Discussion Point: What are some personal benefits you have experienced from maintaining good attendance?

Consequences of Poor Attendance:

- **Disruption**: Poor attendance can disrupt workflow and cause delays in project completion.
- **Increased Workload**: Absences can lead to an increased workload for other team members, causing stress and potential burnout.
- **Negative Impact**: Frequent tardiness or absences can negatively impact your performance reviews and career growth.
  Discussion Point: How can poor attendance affect our team and the overall work environment?

Strategies for Improving Punctuality and Attendance:

- **Plan Ahead**: Plan your commute and daily schedule to ensure you arrive on time.
- **Set Reminders**: Use alarms and reminders to help you stay on track with your schedule.
- **Communicate**: If you anticipate being late or absent, communicate with your supervisor as soon as possible.
  Discussion Point: What strategies do you use to ensure you maintain good attendance and punctuality?

Supporting Each Other:

- **Encouragement**: Encourage and support your coworkers in maintaining good attendance.
- **Team Effort**: Work together to cover for each other when necessary, ensuring that the team's productivity is not affected.

- **Positive Environment**: Foster a positive work environment where everyone feels valued and motivated to be present and on time.
  Discussion Point: How can we support each other in maintaining good attendance and punctuality?

Conclusion
Punctuality and attendance are key to a successful and efficient workplace. By understanding their importance, recognizing the benefits, and implementing strategies to improve, we can create a reliable and productive work environment. Let's commit to being present and on time, supporting each other, and maintaining a positive workplace culture.

BATTERY SAFETY

With the increasing use of battery-powered tools, it's crucial to understand the specific safety measures related to batteries.

1. **Inspection and Maintenance**:
   - Regularly inspect batteries for damage, leaks, or corrosion.
   - Clean battery terminals to ensure good contact.
2. **Charging Safety**:
   - Use only the charger specified by the manufacturer.
   - Charge batteries in a well-ventilated area away from flammable materials.
   - Do not overcharge batteries; follow the manufacturer's guidelines.
3. **Handling and Storage**:
   - Store batteries in a cool, dry place away from direct sunlight and heat sources.
   - Avoid dropping or striking batteries, as this can cause internal damage.
4. **Disposal**:
   - Follow local regulations for battery disposal. Do not throw batteries in the trash (Battery Recycling and Disposal Guidelines, 2025).
   - Use designated recycling facilities for battery disposal.

LITHIUM BATTERY SAFETY

Lithium batteries are commonly used in portable electric tools due to their high energy density and long life. However, they require specific safety precautions:

1. **Inspection and Handling**:
   - Inspect lithium batteries for any signs of damage, swelling, or leaks before use.
   - Handle batteries with care to avoid puncturing or crushing them.
2. **Charging Safety**:
   - Use only chargers designed for lithium batteries.
   - Never leave charging batteries unattended.
   - Charge in a fire-safe area, away from flammable materials.

3. **Storage**:
   - Store lithium batteries at room temperature in a dry place.
   - Avoid storing batteries in extreme temperatures (both hot and cold).
4. **Emergency Procedures**:
   - In case of a battery fire, use a Class D fire extinguisher or sand to extinguish it. Do not use water.
   - If a battery is damaged or leaking, isolate it in a non-flammable container and follow proper disposal procedures.

PPE

When working with or maintaining batteries of any size and specifically batteries which require watering or chemical adding, the use of proper PPE is essential. If you have not been trained how to change batteries or how to maintain them, contact a supervisor first. Your safety is your first priority.

Battery PPE

It could include Aprons, Gloves, Face Shields, etc.

Fire Prevention is at the center of battery safety. By following proper storage, use, and disposal procedures we can mitigate the risk.

Stay Safe, Act Safely.

Discuss

How do you care for batteries at home? Did you know we have battery recycling that you can take advantage of?

BBS—DEFINING BEHAVIOR BASED SAFETY

We have all become acquainted with the term 'see something say something'. Whether travelling through an airport, or taking your kids to the park we have been conditioned to look for dangerous situations; Behavior Based Safety:

That same principle applies to our workplace:

BBS: a process that informs management and employees of the overall safety of the workplace through observations.

The goal is to change unsafe behavior through positive action.

If you see a co-worker not following safety protocol and you choose not to tell your supervisor, you have left the potential for harm to that person and others in the area.

If the situation calls for immediate action such as a spill, advise your co-workers in the area then inform your supervisor.

Perhaps you know of a situation that was not covered during training or a process that changed that does not seem safe, ask questions, tell your supervisor!

Typically, we think of the one-off safety issues, spills, tripping hazards, equipment failure however, safety hazards can occur in seemly benign settings.

- **Good lighting in workspace**
- **Wearing eye protection.**

- **Lifting with ergonomic techniques.**
- **Driving at or below recommended PIV speed.**

Take a moment and think about your workspace. Can you think of a safety concern? Do you know the cause? Do you have a potential resolution?

This is called the ABCs of Safety: *Antecedent, Behavior, and Consequence.*
What is the event?
What is the cause?
How to prevent/resolve and if we don't resolve then what?

When you see something, say something to your supervisor. Safety is everyone's responsibility.

BBS—IT'S WHO WE ARE

Behavior Based Safety or BBS is not just a fancy policy, a reactive safety plan or the latest workplace trend. It's not a culture we talk about once a year or a gimmick we try for a month.

BBS is a way of being.

It is about creating a passion and a purpose to protect and celebrate life, every day, in every task we do. It's about doing the right thing, even when no one is watching. Like picking up a piece of trash, not because it's your job, but because it's the right thing to do.

At every level of our organization, each person plays a role in this living, breathing workplace. BBS thrives when we all take ownership. Let's discuss just how we can make any environment a home for this approach. At the end of the day, Safe 2 Home is everyone's goal.

What your role in BBS looks like:

- Own your role: Know your job inside and out. Understand the tasks, the risks, and the expectations.
- Spot and Avoid Hazards: Stay alert. Identify risks before they come incidents. Protect each other.
- Report Immediately: If something happens, say something. Timely reporting saves lives.
- Inspections: Daily check tools and explore your work area. Report anything concerning. Avoid, looking at past problems. Pick a small area to review in detail daily.
- Be a Safety Hero: Not all heroes wear capes, but they do wear PPE.

Managers and Supervisors have additional duties:

- Provide the Tools. Ensure employees have the right PPE, training, and environmental protection for the job safely, quickly, and efficiently.
- Lead by Example: If employees aren't allowed cell phones on the work floor, keep yours in your pocket.
- Support and Empower: Create an environment where safety concerns are heard, addressed, and celebrated.

Today's challenge

In your immediate departments organize a daily inspection list for each day of the week. Assign a team member to complete these in a rotation. Fresh eyes on the same areas every few days will increase the likelihood of preventing a safety incident.

Finally remember we are all in this together. Whether on the floor, behind a desk, or leading a team, safety is our responsibility. When it comes to safety. We are all First Round Draft Picks. Let's win the day.

BLIND SPOTS AND THE POWER OF THE SPOTTER

Backing up a truck can be one of the most challenging and hazardous maneuvers for drivers. Using a passenger as a spotter can significantly reduce the risk of accidents by providing an extra set of eyes to guide the driver. Effective communication, especially through hand signals, is crucial for safe backing.

Communication with Hand Signals

Clear and consistent hand signals are essential for effective communication between the driver and the spotter. Here is a basic hand signal guide:

- **Stop**: Raise one hand with the palm facing the driver. This signal indicates that the driver should immediately stop the vehicle.
- **Move Back**: Wave one hand backward. This signal tells the driver to start moving the vehicle in reverse.
- **Turn Left**: Extend the left arm straight out to the side. This indicates that the driver should turn the vehicle to the left.
- **Turn Right**: Extend the right arm straight out to the side. This indicates that the driver should turn the vehicle to the right.
- **Slow Down**: Move both hands up and down slowly. This signal tells the driver to reduce the speed of the vehicle.
- **Emergency Stop**: Cross both arms above the head. This is a critical signal indicating that the driver must stop the vehicle immediately due to an emergency.

Common Dangers of Unguided Backing:

1. **Blind Spots**: Large trucks have significant blind spots that can hide obstacles, vehicles, or pedestrians.
2. **Limited Visibility**: Mirrors and cameras may not provide a complete view of the area behind the truck.
3. **Unexpected Obstacles**: Objects or people can suddenly appear in the path of the truck.
4. **Miscommunication**: Without clear signals, the driver may misinterpret the spotter's intentions, leading to accidents.

Overhead and Width Clearance:

- **Overhead Clearance**: Always be aware of the height of your truck and any overhead obstacles such as bridges, power lines, and tree branches. The legal height limit in the United States is generally 13 feet 6 inches (Federal Size Regulations for Commercial Motor Vehicles, 2025).
- **Width Clearance**: Ensure there is enough space on both sides of the truck to avoid hitting objects or vehicles. This includes being mindful of narrow streets, construction zones, and parking lots.

General Safety Precautions:

1. **Pre-Backing Check**: Walk around the truck to identify any potential hazards before backing up.
2. **Use a Spotter**: Always use a spotter when available. The spotter should be positioned where they are always visible to the driver.
3. **Clear Communication**: Establish clear hand signals and ensure both the driver and spotter understand them.
4. **Slow and Steady**: Back up slowly and cautiously, constantly checking mirrors and listening to the spotter.

Discussion Points:

1. **Scenario-Based Questions**: Present hypothetical scenarios involving backing up and asking employees how they would respond.
2. **Personal Experiences**: Encourage employees to share their experiences with backing up and using spotters.
3. **Quiz or Trivia**: Conduct a short quiz on hand signals and backing safety.
4. **Group Activities**: Organize small group discussions to solve safety-related problems.
5. **Guest Speakers**: Invite experts to talk about safe backing practices and the importance of using spotters.

By following these guidelines and engaging in open discussions, we can enhance our safety practices and reduce the risks associated with backing up trucks. Stay vigilant and always prioritize safety!

BLOOD-BORNE PATHOGENS

Blood-borne pathogens are infectious microorganisms present in blood that can cause disease in humans. These pathogens include, but are not limited to, hepatitis B (HBV), hepatitis C (HCV), and human immunodeficiency virus (HIV). Understanding how to protect yourself and others from exposure is crucial in maintaining a safe workplace (29 CFR 1910.1030, 2025).

Key Points:

1. **Understanding Blood Borne Pathogens**:
   - Blood borne pathogens are transmitted through contact with infected human blood and other potentially infectious materials (OPIM).
   - Common modes of transmission include needlestick injuries, cuts from other contaminated sharps, and contact of mucous membranes or broken skin with contaminated blood.
2. **Preventive Measures**:
   - **Use Personal Protective Equipment (PPE)**: Always wear gloves, masks, and eye protection when there is a risk of exposure to blood or OPIM.
   - **Proper Disposal**: Dispose of needles and other sharps in designated sharps disposal containers. Never recap needles.
   - **Hand Hygiene**: Wash hands thoroughly with soap and water after removing gloves or other PPE.
3. **Exposure Control Plan**:
   - **Engineering Controls**: Use devices with safety features designed to prevent injuries, such as self-sheathing needles.
   - **Work Practice Controls**: Implement practices that reduce the likelihood of exposure, such as not eating or drinking in areas where blood or OPIM could be present.
   - **Training and Education**: Regularly train employees on the risks of blood borne pathogens and the proper procedures to follow to minimize exposure.
4. **What to Do in Case of Exposure**:
   - **Immediate Action**: Wash the exposed area with soap and water. If eyes or mucous membranes are exposed, flush with water.
   - **Report the Incident**: Notify your supervisor immediately and seek a medical evaluation.
   - **Follow-Up**: Follow the recommended post-exposure evaluation and follow-up procedures, including necessary blood tests and vaccinations.

Questions to Discuss:

1. What are some common blood borne pathogens we need to be aware of?
2. What personal protective equipment should be used when dealing with potential exposure?
3. How should needles and other sharps be disposed of safely?
4. What steps should you take if you are exposed to blood or infectious substances?
5. Why is it important to report all exposures immediately?

Conclusion

Being informed and vigilant about blood borne pathogens is essential for maintaining a safe and healthy workplace. By following proper procedures and using protective measures, we can significantly reduce the risk of exposure and ensure everyone's safety.

BOOM LIFTS

Boom lifts are essential for reaching high places safely, but their operation requires careful planning and communication. Proper pre-work planning and effective communication can prevent accidents and ensure a smooth workflow.

Key Points:

1. **Pre-Work Planning**:
   - **Site Assessment**: Evaluate the work area for potential hazards such as uneven ground, overhead obstacles, and weather conditions.
   - **Equipment Inspection**: Conduct a thorough inspection of the boom lift, checking for any damage, fluid leaks, tire condition, and proper functioning of controls and safety devices.
   - **Load Limits**: Verify the load limits of the boom lift and ensure that the combined weight of the operator, tools, and materials does not exceed these limits.
   - **Access and Egress**: Plan safe access routes to and from the work area. Ensure that the path is clear of obstacles and that there is enough space for the boom lift to maneuver.
2. **Communication**:
   - **Pre-Work Meetings**: Hold a pre-work meeting with all team members to discuss the day's tasks, potential hazards, and safety protocols.
   - **Hand Signals and Radios**: Use standardized hand signals and radios for clear communication between the operator and ground personnel.
   - **Emergency Procedures**: Ensure everyone knows the emergency procedures, including how to use the emergency lowering mechanism and the location of first aid kits and emergency contact numbers.
   - **Spotters**: Assign a spotter to assist the operator by providing guidance and monitoring for hazards.
3. **Safe Operation**:
   - **Training**: Ensure that all operators are properly trained and certified to use boom lifts.
   - **Stability**: Set up the boom lift on a stable, level surface. Use outriggers if available and ensure they are properly deployed.
   - **Harnesses and Fall Protection**: Always wear a harness and attach it to the designated anchor point on the boom lift.
   - **Weather Conditions**: Avoid operating boom lifts in high winds, rain, or other adverse weather conditions.
4. **Post-Work Procedures**:
   - **Equipment Shutdown**: Properly shut down and secure the boom lift after use.
   - **Inspection and Reporting**: Conduct a post-use inspection and report any issues or damage to your supervisor.
   - **Debriefing**: Hold a debriefing session to discuss what went well and any areas for improvement.

Conclusion
Pre-work planning and effective communication are critical for the safe operation of boom lifts. By assessing the site, inspecting equipment, holding pre-work meetings, and using clear communication methods, we can prevent accidents and ensure a safe working environment. Let's commit to these practices and prioritize safety in all our operations. Additional information regarding specific fall protection requirements can be found in your 1926 manual (29 CFR 1926.453, 2025).

BURN HAZARDS

Burn injuries at home and in the workplace are extremely common. Thousands of people are hospitalized each year due to workplace burns, some incidents are unfortunately fatal. The American Burn Association suggests this trend will continue into 2025 (Association, 2024). Our objective today is to discuss the types of burns and how we can identify dangers before an injury occurs.

Types:
Thermal—These are what most people think of when it comes to burns. These are direct contact related. Meaning contact with fire, hot surfaces, or explosions.

Chemical- Result from physical contact, inhalation, or ingestion of an acid, corrosive, caustic, or alkaline substance. Commonly pictured are household chemicals such as bleach, rubbing alcohol, or mineral spirits. However, gas, diesel, even spray paint can be hazardous.

Electrical- These burns are typically caused by unexpected arc flashes but can also happen when electrical current passes into the body due to direct contact with an energy source. Electrical burns can be both external and internal. Anyone who experiences a significant electrical event should seek medical attention to ensure no hidden damage to tissue is life threatening.

Burn Prevention
The best prevention of thermal burns is to identify potential hazards. Hot Pipes, or surfaces should be protected by professionally installed guards, made of heat safe materials. These items should also be identified by proper safety signage. Individuals working or living near these areas should be instructed regarding the danger. The last line of defense is proper PPE.

Chemical dangers are best prevented by substituting chemicals used for less harsh chemicals whenever possible. Eliminate access to areas where there is higher risk of exposure. Use signage to identify risk. All chemical containers must be properly labeled and suitable for the containment of the chemical being stored. Finally, proper PPE should always be worn when working with chemicals. See the SDS sheet to identify what PPE is required and what dangers exist.

Electrical burns can be prevented by following safe work practices. Inspect all power tools and equipment for frayed or damaged cords, switches, and connections. Do not operate tools where moisture is present. And always wear proper PPE for the task at hand.

Discuss
What exposure to burns do you have while doing your job? How can we help mitigate the risk of that exposure? Remember the various types of burns as you discuss hazards in the workplace.

Together, we can identify, control, and eliminate burn risks. Let's help each other make it **Safe 2 Home** today.

CELL PHONES

Cell phones are an essential part of our daily lives, but their use in a warehouse setting can pose significant safety risks. This talk will outline the rules and guidelines for cell phone use to ensure a safe working environment, especially around moving equipment.

1. **No Cell Phones While Operating Equipment**
   - **Absolute Prohibition**: Do not use cell phones while operating Powered Industrial Vehicles.
   - **Focus on Task**: Your full focus is on operating the equipment safely. Distractions lead to incidents.
2. **Designated Areas for Cell Phone Use**
   - **Use Only in Designated Areas**: Cell phones should only be used in designated break areas or offices.
   - **No Use on the Floor**: Do not use cell phones where equipment is operated.
3. **Emergency Situations**
   - **Emergency Use**: If you need to use your cell phone for an emergency, move to a safe area away from moving equipment and other hazards.
   - **Inform Supervisor**: Notify your supervisor if you need to take an emergency call.
4. **Reporting Hazards**
   - **Report Immediately**: If you see a safety hazard, report it immediately to a supervisor.
   - **Follow Procedures**: Use the proper channels to report hazards and ensure they are addressed promptly.
5. **Personal Responsibility**
   - **Stay Alert**: Always stay alert and aware of your surroundings. Using a cell phone can distract you from potential hazards.
   - **Lead by Example**: Set a good example for your coworkers by adhering to cell phone use policies.

Safety is everyone's responsibility. If you have specific questions about our cell phone policy, please talk to your supervisor.

CLOTHING SAFETY

Working around equipment can be hazardous if proper safety measures are not followed. One of the key aspects of safety is wearing the right clothing. This toolbox talk will cover the importance of proper clothing, and the dangers associated with improper attire when working around equipment. Additional information about PPE requirements can be found in your 1910 Manual (29 CFR 1910.132 PPE, 2025).

Proper Clothing:

1. **Personal Protective Equipment (PPE)**:
   - **Safety Glasses**: Shields your eyes from debris and chemicals.
   - **Gloves**: Protects your hands from cuts, abrasions, and chemicals.
   - **Reinforced Toe Boots**: Protects your feet from heavy objects and punctures.
2. **Clothing**:
   - **Long Sleeves and Pants**: Provides protection against cuts, abrasions, and chemical splashes.
   - **High-Visibility Vests**: Ensures you are visible to others, especially in low-light conditions.
   - **Fitted Clothing**: Avoid loose clothing that can get caught in machinery.
3. **Additional Gear**:
   - **Hearing Protection**: Use earplugs or earmuffs in noisy environments.
   - **Respirators**: Protects against inhaling harmful dust, fumes, and vapors.

Dangers of Improper Clothing:

1. **Loose Clothing**:
   - Can get caught in moving parts of machinery, leading to serious injuries.
   - Increases the risk of tripping and falling.
2. **Long Hair:**
   - Can get caught in moving parts or machinery, leading to serious injuries.
3. **Inadequate Footwear**:
   - Lack of proper footwear can lead to foot injuries from falling objects or stepping on sharp objects.
   - Slippery soles can cause slips and falls.
4. **Lack of Visibility**:
   - Not wearing high-visibility clothing can result in accidents, especially in areas with moving vehicles or equipment.
5. **No PPE**:
   - Increases the risk of injuries from flying debris, chemical splashes, and loud noises.

Discussion

What dangers can you identify in your work environment that could catch loose clothing? Examples Shelving/Racking, Forklift moving parts, Rotating belts, gears, pulleys, shafts, and chains.

What dangers can you identify in our work environment that could be made worse by worn out shoes? Examples, Wet or oily spots, icy conditions, dust covered floors.

Conclusion
Always ensure you are wearing the appropriate clothing and PPE when working around equipment. Proper attire not only protects you from potential hazards but also ensures a safer working environment for everyone. Stay safe!

COLD STRESS

Cold stress occurs when the body can't maintain its normal temperature due to cold exposure. This can lead to serious health issues like hypothermia, frostbite, and trench foot (Cold Stress Guide, 2025).

Recognizing Symptoms:

1. **Hypothermia**:
   - **Early Signs**: Shivering, fatigue, confusion.
   - **Severe Signs**: No shivering, blue skin, slowed pulse.
2. **Frostbite**: Numbness, tingling, white or grayish skin.
3. **Trench Foot**: Reddening, numbness, swelling.

Response to Symptoms:

1. **Hypothermia**: Move to warmth, remove wet clothing, warm the body, seek medical help.
2. **Frostbite**: Warm the area gently, avoid rubbing, seek medical help.
3. **Trench Foot**: Dry and warm the area, seek medical help if needed.

Prevention:

1. **Dress in Layers**: Wear insulating, waterproof clothing.
2. **Stay Dry**: Change wet clothes quickly.
3. **Take Breaks**: Warm up regularly.
4. **Monitor Conditions**: Use the buddy system.

Conclusion
Recognize symptoms, respond quickly, and take preventive measures to stay safe in cold weather. Stay warm and stay safe!

COMPLACENCY DANGERS

Good morning, team. Today, I want to talk about the dangers of complacency. Complacency happens when we become too comfortable and start ignoring potential risks, which can lead to accidents and injuries.

Understanding Complacency:

1. **What is Complacency?**
   - It's when we get too comfortable and stop paying attention to potential dangers.
2. **Why is it Dangerous?**
   - It can lead to accidents because we ignore safety protocols and procedures.

Real-Life Example
Let me share a real-life story to illustrate the dangers of complacency. A construction worker, who had been on the job for many years, became very familiar with his tasks. One day, he decided to skip wearing his safety harness while working on a scaffold because he felt confident nothing would happen. Unfortunately, he lost his balance and fell, resulting in serious injuries. This incident could have been prevented if he had followed safety protocols and not become complacent.

Preventing Complacency:

1. **Stay Alert:**
   - Always be aware of your surroundings and potential hazards.
2. **Follow Safety Protocols:**
   - Stick to all safety procedures, no matter how routine the task may seem.
3. **Peer Accountability:**
   - Look out for each other and remind colleagues to stay vigilant.

When have you been complacent at work? How can you game-ify your work tasks to reduce complacency?

COMPUTER ABUSE

Computers are essential tools in the workplace, but their misuse can lead to serious consequences, including security breaches, decreased productivity, and legal issues. Understanding what constitutes computer abuse and how to prevent it is crucial for maintaining a safe and efficient work environment.

Key Points:

1. **Understanding Computer Abuse**:
   - **Definition**: Computer abuse includes any unauthorized or inappropriate use of company computers and networks. This can range from accessing prohibited websites to downloading unauthorized software.
   - **Examples**:
     - Visiting non-work-related websites during work hours.
     - Downloading or installing unauthorized software.
     - Sharing confidential information without permission.
     - Using company computers for personal gain or illegal activities.

2. **Consequences of Computer Abuse**:
   - **Security Risks**: Unauthorized downloads and website visits can introduce malware and viruses, compromising the security of the company's network.
   - **Productivity Loss**: Time spent on non-work-related activities reduces overall productivity and can affect team performance.
   - **Legal and Ethical Issues**: Misuse of company computers can lead to legal consequences and damage the company's reputation.
3. **Preventing Computer Abuse**:
   - **Clear Policies**: Ensure that there are clear, written policies regarding acceptable computer use. Employees should be aware of these policies and the consequences of violating them.
   - **Monitoring and Enforcement**: Implement monitoring tools to track computer usage and enforce policies consistently.
   - **Training and Awareness**: Provide regular training sessions on the importance of proper computer use and the risks associated with abuse.
4. **Best Practices for Computer Use**:
   - **Use for Work-Related Activities**: Limit computer use to work-related tasks during work hours.
   - **Secure Access**: Use strong passwords and do not share login credentials. Lock your computer when not in use.
   - **Avoid Unauthorized Downloads**: Only download software and files from trusted sources and with proper authorization.
   - **Report Suspicious Activity**: Immediately report any suspicious emails, downloads, or activities to the IT department.
5. **Public Wi-Fi Privileges**:
   - **Security Risks**: Public Wi-Fi networks are often unsecured, making them vulnerable to cyberattacks. Avoid accessing sensitive company information over public Wi-Fi.
   - **Use VPN**: If you must use public Wi-Fi for work purposes, always use a Virtual Private Network (VPN) to encrypt your connection and protect your data.
   - **Avoid Financial Transactions**: Do not conduct financial transactions or access confidential information over public Wi-Fi.
   - **Disconnect When Not in Use**: Disconnect from public Wi-Fi when you are not actively using it to reduce the risk of unauthorized access.

Conclusion

Preventing computer abuse is essential for maintaining a secure and productive environment. By understanding the risks and following best practices, including being cautious with public Wi-Fi, we can protect our company's data and resources. Let's commit to using our computers responsibly and supporting each other in maintaining a safe digital workplace.

COMMUNICATION WITH TEAMMATES

Effective communication among teammates is essential for a safe and productive work environment. Clear and open communication helps prevent misunderstandings, ensures tasks are completed efficiently, and fosters a positive workplace culture.

Key Concepts:

1. **Active Listening**:
   - **Definition**: Fully concentrating, understanding, responding, and remembering what is being said.
   - **Importance**: Ensures that all team members feel heard and valued, reducing the risk of miscommunication.
   - **Practice**: Make eye contact, nod in agreement, and ask clarifying questions.
2. **Clear and Concise Messaging**:
   - **Definition**: Communicating information in a straightforward and easy-to-understand manner.
   - **Importance**: Helps avoid confusion and ensures that everyone is on the same page.
   - **Practice**: Use simple language, be direct, and avoid unnecessary jargon.
3. **Communicating Hazards**:
   - **Definition**: Informing teammates about potential dangers in the work environment.
   - **Importance**: Ensures that everyone is aware of risks and can take appropriate precautions to stay safe.
   - **Practice**: Report hazards immediately, use clear and specific language, and ensure all team members understand the risks.
4. **Feedback and Constructive Criticism**:
   - **Definition**: Providing and receiving feedback in a way that is helpful and aimed at improvement.
   - **Importance**: Encourages continuous improvement and helps address issues before they become bigger problems.
   - **Practice**: Be specific, focus on the issue, not the person, and offer solutions.

Conclusion

Effective communication among teammates is crucial for a safe and efficient workplace. By practicing active listening, clear messaging, communicating hazards, and providing constructive feedback, we can improve our communication skills and work better together. Let's commit to enhancing our communication for a safer and more productive work environment. Stay connected and stay safe!

CREATING POSITIVE ATTITUDES

Introduction

Good morning, team. Today, I want to talk about something that is fundamental to our success and well-being—creating positive attitudes and the importance of leadership. As your _______________, I believe that a positive attitude is the cornerstone of a productive and harmonious workplace. It not only enhances our individual performance but also strengthens our team spirit.

Creating Positive Attitudes:

1. **Mindset Matters:** A positive attitude starts with the right mindset. Embrace challenges as opportunities to learn and grow. Remember, our attitude can influence our outcomes.
2. **Support Each Other:** Encourage and support your colleagues. A simple word of encouragement can make a significant difference in someone's day.
3. **Stay Solution-Focused:** Instead of dwelling on problems, focus on finding solutions. This proactive approach fosters a positive environment.
4. **Celebrate Successes:** No matter how small, celebrate your achievements and those of your teammates. Recognition boosts morale and motivation.

The Importance of Leadership:

1. **Lead by Example:** As leaders, our actions set the tone for the entire team. Demonstrate the behaviors and attitudes you wish to see in others.
2. **Communicate Effectively:** Open and honest communication builds trust and transparency. Make sure to listen as much as you speak.
3. **Empower Your Team:** Trust your team members and empower them to take ownership of their tasks. This builds confidence and fosters a sense of responsibility.
4. **Provide Support:** Be there for your team. Offer guidance, resources, and support to help them succeed.

Team Building Exercise
To reinforce today's talk, let's engage in a team-building exercise called "Positive Feedback Circle."

Instructions:

1. **Form a Circle:** Gather everyone in a circle.
2. **Share Positives:** Each person will take a turn to share something positive about the person to their right. It could be a compliment, a recognition of their hard work, or a positive trait they admire.
3. **Reflect and Appreciate:** After everyone has shared, take a moment to reflect on the positive feedback received. Appreciate the strengths and contributions of your colleagues.

This exercise helps us recognize and appreciate the positive qualities in each other, fostering a supportive and positive work environment.

Thank you for your attention and participation. Let's continue to build a workplace where positivity and strong leadership thrive. Together, we can achieve great things.

DIESEL DANGERS

Diesel fuel is widely used in various industries for powering engines and equipment. While it is less flammable than gasoline, it still poses significant hazards if not handled properly. This talk will cover what diesel fuel is, its chemical makeup, the dangers of exposure, ignition and fuel source risks, medical response for splash hazards, and storage recommendations.

What is Diesel Fuel?

Diesel fuel is a type of liquid fuel derived from crude oil, used primarily in diesel engines. It is composed of a mixture of hydrocarbons and additives to improve performance and reduce emissions.

Chemical Makeup and Dangers of Exposure:

- **Chemical Composition**: Diesel fuel consists of hydrocarbons, including alkanes, cycloalkanes, and aromatic hydrocarbons.
- **Health Risks**: Prolonged exposure to diesel fuel can cause health issues.
    - **Inhalation**: Breathing in diesel vapors can lead to respiratory problems, dizziness, and headaches.
    - **Skin Contact**: Prolonged skin contact can cause irritation, dermatitis, and other skin conditions.
    - **Carcinogenic Risk**: Some components of diesel fuel, such as polycyclic aromatic hydrocarbons (PAHs), are known to be carcinogenic.

Ignition and Fuel Source Dangers:

- **Flammability**: While diesel fuel is less flammable than gasoline, it can still ignite under certain conditions, especially when heated.
- **Explosion Risk**: In confined spaces, diesel vapors can accumulate and create an explosive atmosphere.
- **Precautions**: Handle diesel fuel in well-ventilated areas and keep it away from ignition sources.

Medical Response for Splash Hazards:

- **Eye Contact**: If diesel fuel splashes into the eyes, rinse immediately with plenty of water for at least 15 minutes and seek medical attention.
- **Skin Contact**: If diesel fuel encounters the skin, wash the affected area with soap and water. Remove contaminated clothing and seek medical attention if irritation persists.
- **Inhalation**: Move the person to fresh air immediately. If they have trouble breathing, seek medical attention.

Storage Recommendations:

- **Approved Containers**: Store diesel fuel in approved, clearly labeled containers designed for flammable liquids.

- **Ventilation**: Store diesel fuel in a well-ventilated area away from living spaces and ignition sources.
- **Temperature Control**: Keep diesel fuel containers in a cool, dry place to prevent pressure buildup and leaks.
- **Spill Prevention**: Ensure containers are tightly sealed and stored upright to prevent spills and leaks.

Safety Data Sheet (SDS)
For detailed information on diesel fuel safety, please refer to the Safety Data Sheet (SDS) provided by your supplier. [Link to SDS Sheet]

Interactive Questions:

1. What are common hazards associated with diesel fuel that you have encountered in your work?
2. How do you ensure that diesel fuel is stored properly in your workplace?
3. Can you describe the steps you would take if diesel fuel splashes into your eyes or on your skin?
4. What precautions do you take to prevent diesel fuel from becoming an ignition source?
5. How do you handle and dispose of diesel fuel spills in your work area?

Conclusion
Understanding the hazards associated with diesel fuel and following proper safety practices can prevent accidents and injuries. Always handle diesel fuel with care, use appropriate protective equipment, and follow storage guidelines to ensure a safe working environment. Stay vigilant and report any safety concerns to your supervisor.

DISTRACTED DRIVING

Whether operating a vehicle on the roadway or a forklift in the warehouse distracted driving is a real safety concern.

In fact, more than 2.5 million distracted driving accidents happened on the road last year. Another half a million accidents were reported on construction and work sites. Distracted driving is quickly becoming the number one cause of motor vehicle deaths worldwide. Its official, driving distracted is more dangerous than intoxicated driving!

Discuss
Discuss with your group some examples of distracted driving? Ask them to remember the examples for further discussion.

The Categories:

Manual- When your hands are occupied by something other than the vehicle controls. Such as phone, drink, or food.

Visual-Anything that takes your eyes away from the direction of travel. An interesting sign, a ringing phone, or even a friend waving to you.

Cognitive- Anything that takes your mind away from the task of operating the vehicle. Daydreaming, singing songs, conversations, etc. This multitasking reduces our ability to react to situations.

Discuss
What categories would the examples the team provided fall into? How may it fall into multiple categories?

What can we do to reduce the risk?
We have several policies and recommendations to prevent distracted driving. Those are available in the Distracted Driving Training and further outlined in the Safety Handbook.

But here are a few quick tips:

- Let it wait- Find a safe time/place to check your phone.
- Allow your passengers to return calls or texts for you.
- Preprogram your response. Such as "I am driving, will call you back."
- Plan. Programs in maps, music, and more.

Distracted Driving may prevent someone from making it Safe 2 Home.

DRUGS AND ALCOHOL

Drug and Alcohol or substance abuse issues have been a part of the human condition since the dawn of humankind. It's no wonder we still are challenged by it today. Today let us look at some key points regarding substance abuse at work and what resources are available to you or others.

1. **Understanding the Impact**:
   - Substance abuse can severely affect workplace safety, productivity, and employee well-being. It can lead to accidents, injuries, and decreased performance.
2. **Prevention Measures**:
   - Promote a healthy lifestyle and provide resources for employees to seek help. Implement regular training and awareness programs about the dangers of substance abuse.
3. **Reporting Concerns**:
   - Encourage employees to report any concerns or suspicions regarding substance abuse to their supervisor or HR department. Confidentiality and support should be assured.
4. **Company Policy**:
   - Clearly communicate the company's drug and alcohol policy, including the consequences of violating it. Ensure all employees understand the importance of adhering to these guidelines.

Signs of Substance Abuse:

- **Frequent Tardiness or Unexplained Absences**:
  - Regular lateness or absences can indicate substance abuse issues.
- **Decreased Performance**:
  - Noticeable drop in work quality or productivity
- **Changes in Appearance and Hygiene**:
  - Neglect of personal grooming and hygiene
- **Mood Swings and Erratic Behavior**:
  - Unpredictable mood changes, irritability, or unusual behavior

Employee Resources:

- **Employee Assistance Programs (EAPs)**:
  - Confidential counseling and support services.
- **Health and Wellness Programs**:
  - Resources for maintaining a healthy lifestyle.
- **HR Department**:
  - Guidance on available support and reporting procedures.

Discussion Points:

1. What resources or support systems does our company offer for employees struggling with substance abuse?
2. How can we create a supportive environment that encourages employees to seek help without fear of stigma or retribution?

ELECTRICAL SAFETY—A SHOCKING EXPERIENCE

Working with electricity can be hazardous if proper safety measures are not followed. This talk will cover ten essential safety tips to ensure safe practices when working with electrical systems.

1. **Understand the Hazards**
   - Be aware of the dangers associated with electricity, including electric shock, burns, and fire.
2. **Training and Certification**
   - Ensure that only qualified and trained personnel work on electrical systems. Proper certification is essential.
3. **Use Personal Protective Equipment (PPE)**
   - Wear appropriate PPE such as insulated gloves, safety glasses, and flame-resistant clothing.
4. **Inspect Tools and Equipment**
   - Regularly inspect electrical tools and equipment for damage. Do not use defective tools.

5. **De-energize Before Working**
   - Always de-energize electrical circuits before starting work. Use lockout tagout (LOTO) procedures to ensure circuits remain de-energized.

Discuss

Review LOTO processes. Ask the team to talk through locking out a particular piece of equipment.

6. **Use Proper Tools**
   - Use tools that are rated for electrical work. Insulated tools are essential to prevent accidental contact with live circuits.
7. **Maintain a Safe Distance**
   - Keep a safe distance from live electrical parts. Use barriers and signs to warn others of electrical hazards.
8. **Avoid Water**
   - Never work with electricity in wet or damp conditions. Water increases the risk of electric shock.
9. **Follow Electrical Codes and Standards**
   - Adhere to local electrical codes and standards to ensure safe installation and maintenance practices.
10. **Emergency Procedures**
    - Know the emergency procedures for electrical accidents, including how to shut off power and provide first aid for electric shock.

Bonus

Arc Flash is one of the highest safety concerns when working with electricity. Although it is covered in more depth, in another training. At no time should any unauthorized or untrained person work inside an electrical panel.

Conclusion

By following these ten safety tips, we can reduce the risks associated with working with electricity and create a safer work environment. Always stay vigilant and report any safety concerns to your supervisor.

EMERGENCY RESPONSE

Emergencies can happen at any time, and being prepared is crucial for ensuring everyone's safety. Let us look at several key aspects of emergency response, focusing on injury response and weather events. We'll discuss who to call, what to say, and what to do in these situations. Let's make this an interactive discussion to ensure everyone understands their role in an emergency.

Who to Call

- **Emergency Services**: Dial 911 for any life-threatening emergencies.
- **Supervisor**: Notify your supervisor immediately after calling emergency services.

- **First Aid Team**: Contact the designated first aid responders in your workplace.
- ****Always notify the Safety Team****

Discussion: Who are the first aid responders in our workplace? Does everyone know how to reach them?

What to Say

- **Injury Response**:
  - **Location**: Provide the exact location of the incident.
  - **Nature of Injury**: Describe the injury and the condition of the injured person.
  - **Number of Injured**: State how many people are injured.
  - **Your Contact Information**: Give your name and phone number to answer questions.
- **Weather Events**:
  - **Type of Event**: Specify the type of weather event (e.g., tornado, severe storm).
  - **Location**: Provide the exact location and any specific areas affected.
  - **Immediate Threats**: Mention any immediate threats to safety (e.g., downed power lines, flooding).
  - **Your Contact Information**: Give your name and phone number to answer questions.

Discussion: What information do you think is most important to convey during an emergency call?

What to Do

- **Injury Response**:
  - **Ensure Safety**: Make sure the area is safe for both the injured person and responders.
  - **Provide First Aid**: Administer first aid if you are trained and it is safe to do so.
  - **Stay Calm**: Keep the injured person calm and reassure them that help is on the way.
  - **Wait for Help**: Stay with the injured person until emergency services arrive.

- **Weather Events**:
  - **Seek Shelter**: Move to a designated safe area, such as a storm shelter or interior room away from windows.
  - **Stay Informed**: Listen to weather updates and follow instructions from authorities.

- **Avoid Hazards**: Stay away from downed power lines, flooded areas, and other hazards.
- **Check on Others**: Ensure that all coworkers are accounted for and safe.

Discussion: What are our designated safe areas for different types of weather events? How can we ensure everyone knows where to go?

Conclusion

Being prepared for emergencies can save lives. By knowing who to call, what to say, and what to do, we can respond effectively to injuries and weather events. Let's stay vigilant and ensure we all understand our roles in an emergency.

ENFORCING THE RULES

Safety professionals know that supervisors at all levels are key to policy positively impacting the people it's created to protect. This means enforcing the rules is essential to the success of any safety program.

Why Enforcement Matters:

1. **Prevention**:
   - Consistency is key here. Proper prevention, trains employees and protects them from hazards before an injury can occur.
2. **Safety Culture**:
   - When Safety is part of the culture, everyone wins. Many major injuries could have been prevented by simply following the existing procedure.
3. **Loss and Liability**:
   - Keeping up with the legal requirements is essential. Employees deserve our protection. We as safety leaders have a duty to honor that commitment.
4. **Let the good times roll**:
   - When workers feel safe, they tend to be in a better mood. Which leads to higher productivity.
5. **Costs**:
   - Preventing incidents drives the mod rate down and reduces costs in key categories. Direct Costs, Indirect Costs, Labor, and Production.

Key Responsibilities of Supervisors:

- **Lead by Example**:
  - Supervisors should model safe behavior and adherence to rules.
- **Conduct Regular Inspections**:
  - Regularly inspect the workplace to identify and address potential hazards.
- **Provide Training**:
  - Ensure employees receive proper training on safety procedures and regulations.

- **Enforce Rules Consistently**:
  - Apply rules fairly and consistently to all employees.
- **Encourage Reporting**:
  - Create an environment where employees feel comfortable reporting hazards and near-misses.

Conclusion
Rules and safety regulations are essential to establishing a safe work environment. Supervisors must lead, by example, conduct regular inspections, provide training, and consistently enforce rules to protect employees and promote a culture of safety. Stay vigilant and prioritize safety every day!

EQUIPMENT INSPECTIONS (GENERAL)

Regular equipment inspections are crucial for maintaining a safe and efficient workplace. Inspections help identify potential issues before they become serious problems, ensuring that equipment operates safely and effectively. This talk will cover the importance of equipment inspections, key inspection points, and include open discussion points to ensure everyone understands their role in maintaining equipment safety.

Importance of Equipment Inspections
Regular inspections help prevent accidents, reduce downtime, and extend the lifespan of equipment. They ensure that equipment is safe to use and operates efficiently.

Discussion Point: How do regular equipment inspections benefit our workplace?

Key Inspection Points:

1. **Visual Inspection**: Check for visible signs of damage, wear, or corrosion. Look for cracks, leaks, and loose parts. 2. **Operational Check**: Test the equipment to ensure it operates correctly. Listen to unusual noises and check for smooth operation. 3. **Safety Features**: Ensure all safety features, such as guards and emergency stops, are in place and functioning properly. **Discussion Point**: What specific things do you look for during an equipment inspection?

Reporting and Documentation:

- **Report Issues**: Immediately report any issues or defects found during inspections to your supervisor.
- **Document Findings**: Keep a record of inspection findings, including the date, equipment inspected, and any issues identified.
- **Follow Up**: Ensure that reported issues are addressed and resolved promptly.
- **Discussion Point**: How do we ensure that issues found during inspections are properly documented and addressed?

Preventive Maintenance:

- **Regular Maintenance**: Follow the manufacturer's recommended maintenance schedule to keep equipment in good working condition.
- **Replace Worn Parts**: Replace worn or damaged parts promptly to prevent further damage and ensure safe operation.
- **Clean and Lubricate**: Regularly clean and lubricate equipment to prevent wear and improve performance.
- **Discussion Point**: What preventive maintenance tasks do you perform regularly on your equipment?

Supporting Each Other:

- **Team Effort**: Work together to ensure all equipment is regularly inspected and maintained.
- **Share Knowledge**: Share tips and best practices for equipment inspections with your coworkers.
- **Encourage Reporting**: Encourage everyone to report any issues or concerns they notice with equipment.
- **Discussion Point**: How can we support each other in maintaining equipment safety and reliability?

Conclusion
Regular equipment inspections are essential for a safe and efficient workplace. By understanding their importance, knowing what to look for, and ensuring proper reporting and maintenance, we can prevent accidents and keep our equipment running smoothly. Let's commit to regular inspections and support each other in maintaining equipment safety.

Remember your efforts today to inspect your equipment will prevent your equipment from injuring you tomorrow.

ERGONOMICS: SHOULDERS, KNEES, AND ELBOWS

Ergonomics is hugely important in what we do in every facet of the organization. Today, let us focus on how our Shoulders, Knees, and Elbows are directly affected by poor ergonomics and discuss some things we can do to protect ourselves.

Shoulders:

- Posture: Maintain a neutral posture. Keep your shoulders relaxed and avoid hunching.
- Lifting: Use your legs, not your back or shoulders. Keep the load close to your body.
- Repetitive movements: Take breaks to avoid repetitive strain, stretch your shoulders regularly.

Knees:

- Lifting: Bend your knees, not your back, when lifting. This reduces strain on your lower back and knees. Remember to "hinge" at your joints.
- Standing: If you stand for extended periods, shift your weight between your legs and use anti-fatigue mats if available.
- Kneeling: Use knee pads or a cushion to protect your knees, a piece of Styrofoam sheeting works in a pinch.

Elbows:

- Workstation Setup: Ensure your workstation is set up so your elbows are at a 90-degree angle when typing or using tools.
- Repetitive Tasks: Avoid repetitive elbow movements. Take breaks and stretch your arms.
- Lifting: Keep your elbows close to your body when lifting to reduce strain.

General Tips:

- Stretching: incorporate regular stretching into your routine to keep your muscles flexible.
- Breaks: Take frequent breaks to rest and reset your posture.
- Ergonomic Equipment: Use ergonomic tools like lift tables, rolling carts, and skates to reduce body strain.

Conclusion
By paying attention to our shoulders, knees, and elbows, we can significantly reduce the risk of injury and improve our comfort at work. Remember, proper ergonomics is not just about preventing injuries but also about enhancing our overall productivity and well-being.

EVACUATION PLANNING

An effective evacuation plan is crucial for ensuring the safety of all employees during emergencies such as fires, chemical spills, or natural disasters. Understanding the evacuation procedures and knowing what to do can save lives and minimize injuries.

Key Points:

1. **Understanding the Evacuation Plan**:
   - Familiarize yourself with the workplace/site evacuation plan.
   - Know the primary and secondary evacuation routes.
   - Identify the location of emergency exits, assembly points, and fire alarms.
2. **Roles and Responsibilities**:
   - **Employees**: Follow the evacuation plan, assist others if safe to do so, and report to the designated assembly point.

- **Evacuation Wardens**: Ensure everyone evacuates safely, check designated areas, and report to emergency personnel.
- **Supervisors**: Account for all team members at the assembly point and provide information to emergency responders.

3. **Evacuation Procedures**:
   - **Stay Calm**: Remain calm and do not panic. Follow instructions from supervisors and evacuation wardens.
   - **Use Designated Routes**: Use the primary evacuation route unless it is blocked. If blocked, use the secondary route.
   - **Do Not Use Elevators**: Always use stairs during an evacuation.
   - **Assist Others**: Help those who may need assistance, such as individuals with disabilities or injuries.
   - **Report to Assembly Point**: Once outside, go directly to the designated assembly point and check in with your supervisor.
4. **Emergency Equipment**:
   - Know the location of fire extinguishers, first aid kits, and emergency alarms.
   - Understand how to use fire extinguishers and other emergency equipment if trained to do so.
5. **Post-Evacuation**:
   - Do not re-enter the building until it has been declared safe by emergency personnel.
   - Participate in debriefings and provide feedback on the evacuation process to improve future responses.

Conclusion

Being prepared for an evacuation is essential for ensuring everyone's safety during an emergency. Familiarize yourself with the evacuation plan, know your routes, and follow procedures calmly and efficiently. By doing so, we can protect ourselves and our colleagues in the event of an emergency.

EXTENSION CORD SAFETY

Extension cords are commonly used in many workplaces to provide power to tools and equipment. However, improper use of extension cords can lead to serious hazards, including electrical shocks, fires, and tripping accidents. It's important to remember extension cords are meant to be used as a temporary solution. When not in use, they should be disconnected. If a permanent solution is needed, contact your Facilities Team.

1. **Inspect Before Use**
   - Always inspect extension cords for damage before use. Look for frayed wires, exposed conductors, and damaged plugs.
   - Do not use cords that show signs of wear or damage.

2. **Use the Right Cord for the Job**
   - Ensure the extension cord is rated for the intended use. Check the cord's amperage rating and make sure it matches the power requirements of the equipment.
   - Use outdoor-rated cords for outdoor use.

Discuss

Did you know the length of the cord can impact the wasted or expended energy to the device being powered? When using a cord longer than needed, it can create sufficient power loss. Longer than necessary cords should never be used, while part of the cord remains wrapped up. As electricity passes through the cord, the tight wrap can cause excess heat to build and short out the cord.

3. **Avoid Overloading**
   - Do not overload extension cords by plugging in too many devices. Overloading can cause overheating and increase the risk of fire.
   - Use a cord with a sufficient gauge to handle the electrical load.

Discuss

Have you ever overloaded an extension cord? What happened and how did you prevent a dangerous outcome?

4. **Proper Placement**
   - Place extension cords in a way that minimizes tripping hazards. Avoid running cords across walkways or doorways.
   - Use cord covers or tape to secure cords to the floor if they must cross high-traffic areas.
5. **Keep Cords Dry**
   - Avoid using extension cords in wet or damp conditions. Water increases the risk of electric shock.
   - Store cords in a dry place when not in use.
6. **Avoid Pinching or Crushing**
   - Do not run extension cords through windows, doors, or other openings where they can be pinched or crushed.
   - Avoid placing heavy objects on top of cords.
7. **Disconnect Properly**
   - Always disconnect extension cords by pulling on the plug, not the cord. Pulling on the cord can damage the wires and connections.
   - Unplug cords when not in use to prevent overheating and reduce the risk of fire.
8. **Use GFCI Protection**
   - Use Ground Fault Circuit Interrupter (GFCI) protection when using extension cords in wet or damp locations.
   - GFCIs can help prevent electric shock by shutting off power if a ground fault is detected.

9. **Store Cords Safely**
   - Coil extension cords loosely to prevent kinks and damage. Avoid tight wrapping that can cause internal wire breakage.
   - Store cords in a cool, dry place away from direct sunlight and heat sources.
10. **Follow Manufacturer's Instructions**
    - Always follow the manufacturer's instructions for the proper use and maintenance of extension cords.
    - Replace cords that are damaged or no longer meet safety standards.

Conclusion
By following these safety tips, we can minimize the risks associated with using extension cords and ensure a safer working environment. Always stay vigilant and report any safety concerns to your supervisor.

EYE PROTECTION

Eye protection is essential in the workplace to prevent injuries from dust, debris, chemicals, and other hazards. Protecting your eyes can prevent serious injuries and ensure you maintain your vision and overall safety.

Key Points:

1. **Importance of Eye Protection**:
   - Eye injuries can occur in a split second and can have long-lasting effects.
   - Wearing appropriate eye protection can prevent up to 90% of eye injuries.
2. **Types of Eye Protection**:
   - **Safety Glasses**: For General use, these protect against dust, and light debris.
   - **Goggles**: Provide an extended layer of protection against liquid and gas exposures.
   - **Face Shields**: Used alone or as a compliment to Safety Glasses or Goggles. Typically providing a higher level of impact protection.
   - **Welding/Grinding Helmets**: Provide protection against extreme heat, light, and hot work hazards.
3. **When to Wear Eye Protection**:
   - Always wear eye protection when working with or around hazardous materials.
   - Use eye protection during tasks that involve cutting, grinding, welding, or handling chemicals.
   - Ensure eye protection is worn in designated areas where hazards are present.
4. **Proper Use and Maintenance**:
   - Ensure eye protection fits properly and is comfortable to wear.
   - Regularly inspect eye protection for scratches, cracks, or other damage.
   - Clean eye protection according to the manufacturer's instructions to maintain clear vision.
   - Replace damaged or worn-out eye protection immediately.

5. **Emergency Response**:
   - In case of an eye injury, seek medical attention immediately.
   - If a chemical splash occurs, rinse the eyes with water for at least 15 minutes and seek medical help.

Conclusion

Eye protection is a simple yet crucial measure to prevent serious injuries in the workplace. Always wear the appropriate eye protection for the task at hand and ensure it is well-maintained. By taking these precautions, we can protect our vision and stay safe on the job.

FALL ARREST SYSTEMS

Good morning, everyone! Today, we're going to discuss the importance of fall arrest systems. These systems are crucial for preventing serious injuries when working at heights. We'll focus on two specific groups: order picker operators and maintenance staff. Let's ensure we all understand how to use these systems correctly to stay safe.

Order Picker Operators: Harness and Lifeline Use:

1. **Proper Harness Fit**
   - **Adjustability**: Ensure your harness fits snugly but comfortably. Adjust the straps so they are not too loose or too tight.
   - **Inspection**: Before use, inspect the harness for any signs of wear, damage, or defects. Look for frayed straps, broken buckles, or other issues.
2. **Lifeline Connection**
   - **Anchor Points**: Always connect your lifeline to a secure anchor point. Ensure the anchor point is rated for the load and positioned above your head.
   - **Connection Check**: Double-check that the lifeline is securely attached to both the harness and the anchor point. Ensure there are no knots or tangles.
3. **Movement and Positioning**
   - **Stay Connected**: Always remain connected to the lifeline when working at height. Avoid disconnecting unless necessary and safe to do so.
   - **Safe Movement**: Move slowly and deliberately to avoid sudden jerks that could cause a fall. Keep the lifeline taut but not overly tight.

Maintenance Staff: Fall Arrest System Use on Roofs, Ladders, and Lifts:

1. **Roof Safety**
   - **Edge Awareness**: Always be aware of roof edges and maintain a safe distance. Use guardrails or warning lines if available.
   - **Harness and Lifeline**: Wear a harness and connect to a lifeline anchored to a secure point. Ensure the lifeline allows for movement without slack.
2. **Ladder Safety**
   - **Three Points of Contact**: Maintain three points of contact (two hands and one foot, or two feet and one hand) when climbing ladders.

- **Harness Use**: When working from a ladder, wear a harness and connect to a lifeline or anchor point to prevent falls.

3. **Lift Safety**
   - **Guardrails**: Ensure the lift's guardrails are in place and secure before starting work.
   - **Harness and Anchor**: Wear a harness and connect to the lift's designated anchor point. Do not lean over the guardrails or extend beyond the lift's platform.

Active Discussion Points:

1. **Why is it important to inspect your harness and lifeline before use?**
2. **What should you do if you find damage or defects in your fall arrest equipment?**
3. **How can you ensure your anchor point is secure and properly rated?**
4. **What are the key differences in using fall arrest systems on roofs, ladders, and lifts?**
5. **Can anyone share a personal experience or near-miss involving fall arrest systems? What did you learn from it?**

Conclusion

Using fall arrest systems correctly is essential for preventing serious injuries when working at heights. By ensuring proper fit, secure connections, and safe movement, we can protect ourselves and our colleagues. Let's all commit to following these guidelines and staying vigilant. If you have any questions or need further training, please speak up. Safety is everyone's responsibility!

FATAL ½ INCH, SLIP, TRIP, FALL

A mere one-half-inch change in elevation between surfaces can create a significant trip hazard, often referred to as the "fatal 1/2 inch." This small difference can lead to serious injuries if not properly addressed. Let us discuss the importance of recognizing and mitigating these hazards.

Understanding the Hazard:

- **Trip Hazards**: Even a slight change in elevation can cause someone to trip and fall, leading to potential injuries such as sprains, fractures, or head injuries.
- **Common Locations**: These hazards are often found at door thresholds, uneven flooring, transitions between different types of flooring, and areas with settled concrete or asphalt.

Preventing Trips and Falls:

1. **Regular Inspections**:
   - Conduct regular inspections of walking surfaces to identify and address any changes in elevation.

2. **Marking Hazards**:
   - Use bright paint or tape to mark changes in elevation, making them more visible.
3. **Repairing Uneven Surfaces**:
   - Fix uneven surfaces promptly to eliminate trip hazards.
4. **Proper Lighting**:
   - Ensure adequate lighting in all areas to help people see potential hazards.
5. **Employee Awareness**:
   - Educate employees about the risks associated with minor changes in elevation and encourage them to report hazards.

Discussion Prompts:

1. **Identifying Hazards**:
   - Can you think of any areas in our workplace where a slight change in elevation might pose a trip hazard? How can we address these hazards effectively?
2. **Personal Experiences**:
   - Have you or someone you know ever tripped over a minor change in elevation? What was the outcome, and what could have been done to prevent it?

Conclusion

The "fatal 1/2 inch" is a small but significant hazard that can lead to serious injuries. By staying vigilant, conducting regular inspections, and addressing hazards promptly, we can create a safer work environment for everyone. Stay safe and watch your step!

FATIGUE?

Fatigue is a common issue that can significantly impact safety and productivity in the workplace. Understanding the symptoms, dangers, and ways to reduce fatigue is crucial for maintaining a healthy and efficient work environment.

Discussion Points:

1. **Symptoms of Fatigue**
   - **Physical Symptoms**: Yawning, heavy eyelids, and difficulty keeping eyes open.
   - **Cognitive Symptoms**: Difficulty concentrating, memory lapses, and slower reaction times.
   - **Emotional Symptoms**: Irritability, mood swings, and decreased motivation.
   - **Discussion**: Have you noticed any of these symptoms in yourself or coworkers? How do you usually address them?
2. **Dangers of Fatigue**
   - **Accidents and Injuries**: Fatigue can lead to decreased alertness and slower reaction times, increasing the risk of accidents and injuries.

- **Reduced Productivity**: Tired employees are less efficient and make more errors, impacting overall productivity.
- **Health Issues**: Chronic fatigue can lead to long-term health problems, such as cardiovascular disease and mental health issues.
- **Discussion**: Can you share any experiences where fatigue affected your work or safety? What were the consequences?

3. **Suggestions for Reducing Fatigue**
   - **Adequate Sleep**: Aim for 7-9 hours of sleep per night to ensure you are well-rested.
   - **Healthy Diet**: Eat balanced meals and stay hydrated to maintain energy levels.
   - **Regular Breaks**: Take short breaks throughout the day to rest and recharge.
   - **Exercise**: Regular physical activity can improve energy levels and overall well-being.
   - **Stress Management**: Practice stress-reducing techniques such as deep breathing, meditation, or talking to a friend.
   - **Work Schedule**: Try to maintain a consistent work schedule and avoid excessive overtime.
   - **Discussion**: What strategies do you use to combat fatigue? Are there any additional suggestions you think could help?

Conclusion

Fatigue is a serious issue that can affect both safety and productivity. By recognizing the symptoms, understanding the dangers, and implementing strategies to reduce fatigue, we can create a healthier and safer work environment.

FIRE FIGHTING

Generally, firefighting is a task best left to professionals in fancy coats and hats. However, our quick reaction to a small fire can prevent fires from growing and endangering people and property.

Fighting Fire with a Fire Extinguisher

We always encourage firefighting only if: The fire is smaller than a standard wastepaper basket, and there is immediate access to an exit. When fighting a fire, we always want to keep our back towards our exit route.

If the fire is fightable, we observe the PASS Method:

P Pull the pin.
A Aim at the base of the fire.
S Squeeze the trigger.
S Sweep from left to right.

Fire Extinguisher Maintenance

Make sure fire extinguishers are inspected monthly by onsite staff. Inspection should consist of checking the visual appearance of the equipment, ensuring the pressure valve shows fully charged, typically the gauge will be in the green. Annual inspections are completed by professional service technicians.

Discuss and Demonstrate
As a group, demonstrate how to use a fire extinguisher using the PASS method. (*Do not discharge the extinguisher.*) Discuss the process and answer questions.

FIRE EXTINGUISHER USE AND INSPECTION

Introduction
Fire extinguishers are crucial tools for preventing small fires from escalating. However, they are not suitable for large or rapidly spreading fires. The primary focus during a fire should be personal safety and contacting the appropriate authorities. It's essential to understand how to use fire extinguishers and recognize their limitations.

Using a Fire Extinguisher: The P.A.S.S. Method
The P.A.S.S. method is an easy way to remember the steps for using a fire extinguisher:

- **P—Pull**: Pull the pin. Hold the extinguisher away from you and release the locking mechanism.
- **A—Aim**: Aim the nozzle at the base of the fire, not the flames.
- **S—Squeeze**: Squeeze the lever slowly and evenly to release the extinguishing agent.
- **S—Sweep**: Sweep the nozzle from side to side at the base of the fire until it is extinguished.

Fire Extinguisher Range:

- **Range**: A dry chemical fire extinguisher (common red "ABC" type) can reach between 5 and 20 feet. Familiarize yourself with the models in your work area and their effective range.
- **Duration**: A 10lb to 20lb dry chemical extinguisher lasts between 10 to 25 seconds, depending on the model and weight.
- **Size of Fire**: Fire extinguishers are designed for small fires, typically no larger than a small trash can.

Fire Extinguisher Inspection Tips:

- **Monthly Checks**: Inspect extinguishers every 30 days. Conduct a formal annual inspection and document it.
- **Pressure Check**: Ensure the pressure gauge arrow is in the green zone. If it's in the red, tag the extinguisher out of service until recharged.
- **Pin and Condition**: Verify the pin is in place to prevent accidental discharge. Check for rust and ensure the label is readable.

Summary
Knowing the location of fire extinguishers is not enough. Ensure you understand how to use them properly and regularly inspect them to maintain their condition.

Discussion Points:

1. **Why is it important to prioritize personal safety over property during a fire?**
2. **Can you demonstrate the P.A.S.S. method for using a fire extinguisher?**
3. **What are the key points to check during a fire extinguisher inspection?**

FIRE EXTINGUISHER TYPES AND USES

Fire extinguishers are essential safety tools in any workplace. Understanding the different types of fire extinguishers and their specific uses can help prevent small fires from becoming major incidents. This talk will cover the main types of fire extinguishers and the types of fires they are designed to combat.

Types of Fire Extinguishers:

1. **Water (Class A) Extinguishers**
   - **Use**: Effective against ordinary combustibles such as wood, paper, and cloth.
   - **Not For**: Do not use for electrical fires or flammable liquids.
2. **Foam (Class A and B) Extinguishers**
   - **Use**: Suitable for both Class A fires (ordinary combustibles) and Class B fires (flammable liquids like gasoline, oil, and paint).
   - **Not For**: Not recommended for electrical fires.
3. **Dry Powder (Class A, B, and C) Extinguishers**
   - **Use**: Versatile and effective in Class A, B, and C fires (electrical fires).
   - **Special Types**: Some dry powder extinguishers are designed specifically for Class D fires (flammable metals).
4. **Carbon Dioxide (CO_2) (Class B and C) Extinguishers**
   - **Use**: Ideal for electrical fires and flammable liquid fires.
   - **Not For**: Not effective on Class A fires (ordinary combustibles).
5. **Wet Chemical (Class K) Extinguishers**
   - **Use**: Specifically designed for kitchen fires involving cooking oils and fats.
   - **Also Effective**: Can be used on Class A fires.

Conclusion
Understanding the different types of fire extinguishers and their uses is crucial for workplace safety. By using the correct extinguisher, you can effectively manage small fires and prevent them from escalating. Always stay vigilant and report any safety concerns to your supervisor.

FIRE PREVENTION

Fires are almost always preventable, and almost always devastating to people, property, and businesses. So why do they occur so often? Complacency is the number one suspect. When we overlook a danger, or make temporary exceptions to rules, fail to maintain equipment or inspect it, disaster strikes. Fire is avoidable by mitigating heat levels, and removing ignition source risk, as well as containing sources of combustibles in negative pressure spaces, away from sources of ignition.

The Triangle

Fire requires 3 main components. Oxygen, Fuel, and Heat. Removing one component greatly reduces the risk of fire.

Oxygen removal will quench most fires. However, some systems will draw in air, which is 21% oxygen. Induction Fans, air compressors, and air canisters can provide the oxygen sources required for a fire to start and grown. Controlling these systems and isolating these sources will prevent fires and fire growth.

Fuels such as gasoline, diesel, and solvents are commonly thought of, but dust, wood shavings, and other fine materials can also be substantial sources of fuel for fires. Some of the worst fires in US history have been started by statically charged corn chaff. The best prevention is to isolate fuel sources by properly storing them in flame proof cabinets and keeping spills contained and cleaned up.

Heat is the final component. Sources of ignition such as outlets, static, open flames, or sparks are just a few common causes. Friction over time can create enough heat to ignite nearby fuel sources. It is a common occurrence that vehicle brake pad heat, can and will ignite greases, and fluids that build up near the wheel assemblies.

Final Thoughts

Fire prevention is most effectively achieved through dedicated site walks, inspections, and process review. Proper maintenance can eliminate friction, static, and other ignition sources. Together we can eliminate most fire dangers.

FIRST AID BASICS

First aid is the immediate assistance given to someone who is injured or suddenly ill before professional medical help arrives. Knowing basic first aid can save lives, reduce recovery time, and prevent minor injuries from becoming major ones. As anything, First Aid should only be completed if you feel comfortable and more advanced first aid such as CPR or the Heimlich should only be attempted if you are trained in those techniques.

Key Points:

1. **Assess the Situation**: Ensure the area is safe for you and the injured person. Look for any hazards that could cause further injury.
2. **Call for Help**: Dial emergency services immediately if the injury is serious. Provide clear information about the location and nature of the injury.

3. **Basic First Aid Steps**:
    - **CPR (Cardiopulmonary Resuscitation)**: If the person is not breathing, start CPR. Remember the ratio of 30 chest compressions to 2 rescue breaths. Or utilize a provided AED.
    - **Bleeding**: Apply pressure to the wound with a clean cloth to stop the bleeding. Elevate the injured area if possible.
    - **Burns**: Cool the burn under running water for at least 10 minutes. Do not apply ice or ointments.
    - **Choking**: Perform the Heimlich maneuver if someone is choking. Stand behind them, wrap your arms around their waist, and give quick, upward thrusts.
    - **Fractures**: Immobilize the injured area and avoid moving the person unless necessary.

First Aid Kit Essentials:

- Adhesive bandages
- Sterile gauze pads
- Adhesive tape
- Antiseptic wipes
- Scissors
- Tweezers
- CPR mask
- Gloves
- Emergency contact numbers

Questions to Discuss:

1. **Where are the first aid kits located in our workplace?**
2. **Who is trained in first aid and CPR in our team?**
3. **What should you do if you find someone unconscious?**
4. **How do you handle a situation where someone is bleeding heavily?**
5. **What steps should you take if you encounter a burn injury?**
6. **If you need more First Aid Supplies, who should you contact?**

Conclusion
Understanding and practicing basic first aid can make a significant difference in an emergency. Always stay calm, act quickly, and ensure you know the location of first aid kits and emergency contacts.

FOOTWEAR

Wearing good shoes is incredibly important for creating a strong base for lifting and moving. Proper footwear is essential for maintaining balance, preventing injuries, and enhancing overall safety and efficiency on the job.

Benefits of Good Footwear:

1. Enhanced Stability:
   a. Good footwear provides a solid and stable base, which is crucial when lifting and moving heavy objects.
   b. Stability helps you maintain proper posture and balance, reducing risk of slip, trip, and fall.
2. Improved Support:
   a. Quality footwear offers adequate arch and ankle support, which is essential for preventing strain and injuries.
   b. Proper support helps distribute weight evenly across your feet, reducing the risk of fatigue and discomfort.
3. Increased Traction:
   a. Shoes with non-slip soles provide better grip on various surfaces, preventing accidental slips, trips, and falls.
   b. Enhanced traction is especially important in environments where floors may become wet or oily.
4. Protection from Hazards:
   a. Proper footwear protects your feet from potential hazards such as falling objects, sharp materials, and electrical risks.
   b. Steel-toed boots or shoes with reinforced toes can prevent injuries.
5. Comfort and Endurance:
   a. Comfortable footwear reduces pressure points and friction, allowing you to work longer without discomfort.
   b. Reduce fatigue means you can maintain focus and productivity throughout your shift.

Footwear Choice:

- Fit and Comfort:
  - Ensure your shoes fit well: Toes should have some movement, but shoes should be snug in the ankle and ball of the foot.
  - Look for features such as cushioned insoles and breathable materials for added comfort.
- Task appropriateness
  - Select footwear that is suitable for the specific tasks you're doing in your work environment.
    - Slip resistance on wet or dusty floors
    - Reinforced toes to protect against falling objects
- Regular maintenance
  - Inspect your footwear regularly for signs of wear or tear. Replace shoes that are damaged or no longer provide the necessary features or protections your work requires.
  - Keep your shoes clean and dry to maintain their effectiveness and longevity.

Conclusion

Wearing good footwear is a simple yet crucial step in ensuring safety and efficiency when lifting and moving. By choosing the right shoes and maintaining them properly, we can create a strong base that supports our work and protects us from injuries. Let's make sure we prioritize our footwear and take care of our feet.

FORKLIFT SAFETY

Good morning, everyone! Today, we're going to talk about forklift safety in our warehouse. Forklifts are essential for our operations, but they can also be dangerous if not used properly. By following safety guidelines and being aware of our surroundings, we can prevent accidents and injuries.

Discussion Points:

1. **Pre-Operational Checks**
   - **Inspection**: Before using a forklift, conduct a thorough inspection. Check the tires, brakes, steering, lights, and horn. Ensure the forks are in good condition and properly aligned.
   - **Fluid Levels**: Check the fuel, oil, and hydraulic fluid levels. Top them up if necessary.
   - **Battery**: If using an electric forklift, ensure the battery is fully charged and securely connected.
2. **Safe Operation**
   - **Seatbelt Use**: Always wear your seatbelt while operating the forklift.
   - **Speed Control**: Operate the forklift at a safe speed, especially when turning corners or navigating through tight spaces.
   - **Load Handling**: Ensure loads are stable and secure before lifting. Do not exceed the forklift's load capacity.
   - **Visibility**: Make sure you have a clear view of your path. Use mirrors and look in the direction of travel. If your view is obstructed, drive in reverse or use a spotter.
3. **Workplace Awareness**
   - **Pedestrian Safety**: Be aware of pedestrians in the warehouse. Use the horn to alert others of your presence, especially at intersections and blind spots.
   - **Aisle Clearance**: Keep aisles clear of obstructions. Ensure there is enough space for the forklift to maneuver safely.
   - **Floor Conditions**: Be mindful of floor conditions. Avoid wet or slippery areas and report any spills or hazards immediately.
4. **Loading and Unloading**
   - **Proper Stacking**: Stack loads evenly and securely. Do not stack loads too high, as this can affect stability.
   - **Fork Positioning**: Position the forks properly under the load. Tilt the mast back slightly to secure the load before lifting.
   - **Loading Docks**: Be cautious when working near loading docks. Ensure the dock plate is secure and can support the weight of the forklift and load.

5. **Parking and Shutdown**
   - **Safe Parking**: Park the forklift in a designated area. Lower the forks to the ground, set the parking brake, and turn off the engine.
   - **Key Removal**: Remove the key when the forklift is not in use to prevent unauthorized operation.

Active Discussion Points:

1. **Why is it important to conduct pre-operational checks on a forklift?**
2. **What are some common hazards you should be aware of while operating a forklift in the warehouse?**
3. **How can you ensure pedestrian safety while using a forklift?**
4. **Can anyone share a near-miss or incident involving a forklift? What did you learn from it?**
5. **What steps should you take if you encounter an obstruction or hazard in your path?**

Conclusion
Forklift safety is crucial for preventing accidents and ensuring a smooth workflow in our warehouse. By following these guidelines and staying vigilant, we can protect ourselves and our colleagues. Let's all commit to operating forklifts safely and responsibly. If you have any questions or need further training, please speak up. Safety is everyone's responsibility!

GASOLINE DANGERS

Gasoline is a common fuel used in many workplaces, but it poses significant hazards if not handled properly. This talk will cover what gasoline is, its chemical makeup, the dangers of benzene, ignition and fuel source risks, medical response for splash hazards, and storage recommendations.

What Is Gasoline?
Gasoline is a highly flammable liquid used primarily as a fuel in internal combustion engines. It is a complex mixture of hydrocarbons, including additives to enhance performance and reduce emissions.

Chemical Makeup and Dangers of Benzene

- **Chemical Composition**: Gasoline is composed of various hydrocarbons, including alkanes, cycloalkanes, and aromatic hydrocarbons.
- **Benzene**: One of the hazardous components of gasoline is benzene, a known carcinogen.
  - **Health Risks**: Benzene exposure can cause serious health issues, including leukemia and other blood disorders.

- **Inhalation**: Breathing in benzene vapors can lead to dizziness, headaches, and respiratory problems.
- **Skin Contact**: Prolonged skin contact with benzene can cause irritation and dermatitis.

Ignition and Fuel Source Dangers:

- **Flammability**: Gasoline vapors are highly flammable and can ignite easily from sparks, open flames, or hot surfaces.
- **Explosion Risk**: In confined spaces, gasoline vapors can accumulate and create an explosive atmosphere.
- **Precautions**: Always handle gasoline in well-ventilated areas and keep it away from ignition sources.

Medical Response for Splash Hazards:

- **Eye Contact**: If gasoline splashes into the eyes, rinse immediately with plenty of water for at least 15 minutes and seek medical attention.
- **Skin Contact**: If gasoline encounters the skin, wash the affected area with soap and water. Remove contaminated clothing and seek medical attention if irritation persists.
- **Inhalation**: Move the person to fresh air immediately. If they have trouble breathing, seek medical attention.

Storage Recommendations:

- **Approved Containers**: Store gasoline in approved, clearly labeled containers designed for flammable liquids.
- **Ventilation**: Store gasoline in a well-ventilated area away from living spaces and ignition sources.
- **Temperature Control**: Keep gasoline containers in a cool, dry place to prevent pressure buildup and leaks.
- **Spill Prevention**: Ensure containers are tightly sealed and stored upright to prevent spills and leaks.

Safety Data Sheet (SDS)
For detailed information on gasoline safety, please refer to the Safety Data Sheet (SDS) provided by your supplier.

Conclusion
Understanding the hazards associated with gasoline and following proper safety practices can prevent accidents and injuries. Always handle gasoline with care, use appropriate protective equipment, and follow storage guidelines to ensure a safe working environment. Stay vigilant and report any safety concerns to your supervisor.

GET A GRIP!

Gloves enhance our lifting and carrying capacity more than you know. Gloves are a crucial part of our PPE and can significantly impact on our safety and efficiency on the job.

Why Gloves Matter:

1. **Enhanced Grip:**
   a. Gloves provide a better grip on objects, reducing the risk of slips, drops, and sudden reactionary movements.
   b. A secure grip means you can lift and carry more confidently with more control. They reduce strain on your hands and arms.
2. **Protection from injury:**
   a. Gloves protect your hands from cuts, abrasions, and punctures. This is vital when handling sharp or rough materials.
   b. They also shield your skin from harmful substances and extreme temperatures, preventing burns and chemical injuries.
3. **Improved Comfort:**
   a. Wearing gloves can reduce the pressure and friction on your hands. Making it more comfortable to lift and carry items for extended periods.
   b. This comfort can lead to better endurance and less fatigue throughout your day.
4. **Increased Load Capacity:**
   a. With a better grip and more comfort, you can handle heavier loads more safely.
   b. Properly fitted gloves distribute weight evenly across your hand, reducing injury.

Best Practice:

1. Choose the right gloves:
   a. Gloves are designed to meet the needs of each task. They may have other benefits and downfalls.
      i. Downfalls may include warm or sweaty hands. Having multiple pairs to change in and out of during the day can help with this.
   b. Make sure they fit snuggly, but comfortably, particularly around the wrist. Proper blood flow and movement are essential.

Conclusion: Using gloves is a simple yet effective way to improve your lifting and carrying capacity while protecting your hands from injuries. By choosing the right gloves and using them correctly, we can enhance our safety and efficiency on the job. Let's make sure we always wear our gloves and take care of them to keep our hands safe and strong.

HAND PROTECTION

Hand protection is crucial in the workplace to prevent injuries and ensure safety. Using gloves, especially when lifting, can provide significant benefits. Today we will cover the importance of hand protection, the types of gloves available, and the specific benefits of using gloves for lifting tasks.

Importance of Hand Protection
Hands are one of the most vulnerable parts of the body in many work environments. Protecting them can prevent cuts, abrasions, burns, and other injuries.

Discussion Point: What types of hand injuries have you seen or experienced in our workplace?

Types of Gloves:

- **Cut-Resistant Gloves**: Protect against sharp objects and materials.
- **Heat-Resistant Gloves**: Provide protection from high temperatures and burns.
- **Chemical-Resistant Gloves**: Shield hands from harmful chemicals and substances.
- **General-Purpose Gloves**: Offer basic protection for a variety of tasks, including lifting.

Always use the correct hand protection for the task. If unsure, review the SDS for the chemical/substance. If there is no chemical concern, ask a supervisor which glove type might work best.

Discussion Point: What types of gloves do you use most frequently, and for what tasks?

Fit
Gloves should fit snug but not reduce circulation. Sizing can be determined best by ensuring the glove material contacts the space between fingers, without causing fingertip discomfort.

Benefits of Using Gloves for Lifting:

- **Improved Grip**: Gloves can enhance your grip on objects, reducing the risk of dropping items and causing injury.
- **Hand Protection**: Gloves protect your hands from cuts, abrasions, and blisters that can occur during lifting.
- **Reduced Fatigue**: Padded gloves can reduce hand fatigue by providing cushioning and support.
- **Temperature Control**: Gloves can protect your hands from extreme temperatures, whether hot or cold, when handling materials.

Discussion Point: How have gloves helped you with lifting tasks? Can you share any specific examples?

Proper Use and Maintenance of Gloves:

- **Correct Fit**: Ensure gloves fit properly to provide maximum protection and comfort.
- **Regular Inspection**: Check gloves regularly for signs of wear and tear. Replace damaged gloves immediately.
- **Cleanliness**: Keep gloves clean to maintain their protective qualities and prevent skin irritation.

Discussion Point: What are some best practices you follow to ensure your gloves are in good condition?

Conclusion
Hand protection is essential for preventing injuries and ensuring safety in the workplace. Using gloves, especially for lifting tasks, provides numerous benefits, including improved grip, protection, reduced fatigue, and temperature control. Let's commit to using the right gloves for the job and maintain them properly to keep our hands safe.

HAND TOOL SAFETY

Every year we suffer plenty of hand tool injuries. These injuries occur while using box knives, screw drivers, pliers, etc. Many of these stems from improper use or a tool not designed for the job or a damaged tool.

Discuss
What are some common Hand Tools and how might they be dangerous? Hammers, screwdrivers, box knives, chisels, wrenches, etc.

Prevention through Inspection Examples
Hammers—These should not be damaged at the head or the handle. If the hammer is altered or repaired in any way it should be disposed of and replaced. The face of the hammer as well as the claw or pummel should also be inspected for any damage.

Screwdrivers—There should be no damage to the shank, tip, or grip. If damaged or worn, it is time to replace. Screwdrivers should not be used as punches, or pry bars.

Chisels—A dull chisel is a broken chisel. Chisels can be sharpened by professionals. However, they should be disposed of if cracks or bends begin to develop.

Wrenches—There should be no damage or indication of heat stretching. Wrenches with worn or broken teeth should be discarded. Wrenches with adjustable jaws should be inspected and should hold position unless acted upon.

General Hand Tool Best Practice:

- Keep Tools clean.
- Do not modify tools, use the right tool.

- Secure all hand tools and store them properly.
- Use tools how they are designed to be used.
- When cutting, striking, or manipulating material with a tool, always direct energy away from your body.

Challenge

Today look at hand tools in your area. Turn any tools into your supervisor that are broken or need to be replaced. If you find a tool, you're not familiar with or would like further instruction on how to use it safely, please talk to your supervisor.

Together, we can all work to improve our chances of making it Safe 2 Home.

HANDLING HAZARDOUS MATERIALS

Definition

Hazardous materials (HazMat) are substances that can pose a risk to health, safety, or the environment. These include chemicals that are toxic, flammable, corrosive, reactive, or otherwise harmful (29 CFR 1910.120 Hazardous Waste Operations and Emergency Response, 2025).

Types of Exposure

Exposure to hazardous materials can occur in several ways:

- **Inhalation**: Breathing in vapors, fumes, dust, or gases.
- **Skin Contact**: Direct contact with the skin, leading to absorption.
- **Ingestion**: Swallowing hazardous substances, often accidentally.
- **Injection**: Penetration through the skin by sharp objects contaminated with hazardous materials.

Safety Data Sheets (SDS)

For detailed information on specific hazardous materials, please refer to our Safety Data Sheets (SDS) platform.

General Safety Precautions:

1. **Identify Hazards**: Know which materials in your workplace are hazardous.
2. **Proper Training**: Ensure all employees are trained in handling hazardous materials.
3. **Use PPE**: Always wear appropriate personal protective equipment (PPE).
4. **Labeling**: Ensure all containers are properly labeled.
5. **Storage**: Store hazardous materials in appropriate, labeled containers in a cool, dry, and ventilated area.
6. **Emergency Procedures**: Be familiar with emergency procedures, including the use of eyewash stations and showers.

Communication Tips:

- **Regular Meetings**: Conduct regular safety meetings to discuss hazards and safety practices.
- **Clear Labels and Signs**: Use clear and consistent labeling and signage for hazardous materials.
- **Report Incidents**: Encourage prompt reporting of any incidents or near-misses involving hazardous materials.
- **Accessible SDS**: Ensure that SDS are easily accessible to all employees.
- **Feedback Loop**: Create a feedback loop where employees can voice concerns and suggestions regarding hazardous materials.

Ideas for Creating Discussion:

1. **Scenario-Based Questions**: Present hypothetical scenarios involving hazardous materials and ask employees how they would respond. This encourages critical thinking and practical application of safety protocols.
2. **Personal Experiences**: Invite employees to share their own experiences with hazardous materials. This can highlight real-world challenges and solutions.
3. **Quiz or Trivia**: Conduct a short quiz or trivia session on hazardous materials safety. This can be a fun way to reinforce knowledge.
4. **Group Activities**: Organize small group activities where employees can discuss and solve safety-related problems together.
5. **Guest Speakers**: Invite experts to talk about specific hazardous materials and their safe handling.
6. **Visual Aids**: Use videos, posters, or demonstrations to illustrate the dangers and proper handling of hazardous materials.
7. **Open Forum**: Allow time for an open forum where employees can ask questions and discuss concerns about hazardous materials.

By incorporating these discussion ideas, you can make your toolbox talks more engaging and effective, ensuring that everyone is well-informed and prepared to handle hazardous materials safely.

HAZARDOUS WEATHER

Inclement weather can pose significant risks to our safety and operations. It's crucial to be prepared and know how to respond effectively. Today, we'll discuss key aspects of inclement weather safety, including where to go, communication, first aid, and emergency response. Remember to follow the Emergency Action Plan.

Discussion Points:

1. **Where to Go During Inclement Weather**
    - **Designated Safe Areas**: Identify and familiarize yourself with the designated safe areas in your workplace, such as storm shelters or interior rooms away from windows.

- **Evacuation Routes**: Know the evacuation routes and assembly points in case of severe weather like tornadoes or hurricanes.
- **Emergency Bag**: Ensure you take the emergency bag with you, which should contain essential supplies like first aid kits, water, and emergency contact information.
- **Group Captains**: Designate group captains by seniority to lead and coordinate the evacuation process.
- **Discussion**: Do you know where the nearest safe area is? How can we ensure everyone is aware of these locations?

2. **Communication During Inclement Weather**
    - **Emergency Alerts**: Ensure you are signed up for any workplace emergency alert systems and know how to receive updates.
    - **Communication Plan**: Establish a communication plan with your team, including who to contact and how to stay informed during severe weather.
    - **Discussion**: What methods of communication do you find most effective during emergencies? How can we improve our current communication plan?
3. **First Aid During Inclement Weather**
    - **First Aid Kits**: Know the locations of first aid kits and ensure they are well-stocked and accessible.
    - **Basic First Aid Skills**: Familiarize yourself with basic first aid skills, such as treating cuts, bruises, and minor injuries that may occur during an emergency.
    - **Discussion**: Have you ever had to use first aid during an emergency? What additional training or resources would be helpful?
4. **Emergency Response**
    - **Emergency Procedures**: Review the emergency procedures for different types of inclement weather, such as thunderstorms, floods, and extreme heat or cold.
    - **Role Assignments**: Understand your role and responsibilities during an emergency, whether it's assisting with evacuation, providing first aid, or communicating with emergency services.
    - **Discussion**: What are your concerns about our current emergency response plan? How can we address them to ensure everyone feels prepared?

Conclusion

Being prepared for inclement weather is essential for our safety and well-being. By knowing where to go, maintaining clear communication, being equipped with first aid knowledge, and understanding our emergency response roles, we can effectively handle any weather-related challenges.

HAZARD COMMUNICATION

Hazard Communication, also known as HazCom, ensures that employees are informed about the hazards associated with chemicals and dangers may be exposed to. (29 CFR 1910.1200 Hazard Communication Standard, 2025). This talk will cover the key elements of Hazard Communication and include an interactive activity.

Discussion Points:

1. **Understanding Hazard Communication**
   - **Purpose**: To inform employees about chemical hazards and protective measures.
   - **Key Components**: Labels, Safety Data Sheets (SDS), and employee training.
2. **Labels**
   - **Information on Labels**: Product identifier, signal word, hazard statements, precautionary statements, and pictograms.
   - **Signal Words**: "Danger" for severe hazards and "Warning" for less severe hazards.
   - **Pictograms**: Visual symbols that indicate specific hazards (e.g., flammable, toxic).
3. **Safety Data Sheets (SDS)**
   - **Sections of an SDS**: Identification, hazard identification, composition, first-aid measures, fire-fighting measures, accidental release measures, handling and storage, exposure controls/personal protection, physical and chemical properties, stability and reactivity, toxicological information, and more.
   - **Accessing SDS**: Know where SDS are stored and how to access them online.
4. **Employee Training**
   - **Training Requirements**: Employees must be trained on the HazCom program, how to read labels and SDS, and the hazards of chemicals they may be exposed to.
   - **Emergency Procedures**: Understanding what to do in case of a chemical spill or exposure.

Example Activity: SDS Scavenger Hunt

Objective: To familiarize employees with locating and understanding Safety Data Sheets.

Materials Needed: Copies of SDS for various chemicals used in your workplace, a list of questions related to the SDS content.

Instructions:

1. **Divide the group into teams**.
2. **Distribute SDS copies** to each team.
3. **Provide a list of questions** that can be answered using the SDS (e.g., "What is the first-aid measure for skin contact with this chemical?" or "What is the flashpoint of this chemical?").

4. **Set a time limit** for teams to find the answers.
5. **Review the answers** together and discuss any questions or clarifications needed.

Conclusion

Understanding and properly using Hazard Communication tools like labels and SDS can significantly reduce the risk of chemical-related injuries and illnesses. Regular training and activities like the SDS scavenger hunt help reinforce this knowledge.

HAZARD COMMUNICATION (ENVIRONMENTAL)

Effective hazard communication is essential for maintaining a safe workplace (29 CFR 1910.1200 Hazard Communication Standard, 2025). This talk will cover the communication of environmental, process, and equipment hazards, ensuring everyone understands the risks and how to mitigate them.

Discussion Points:

1. **Environmental Hazards**
   - **Chemical Spills**: Understand the procedures for reporting and cleaning up spills. Use appropriate PPE and follow the SDS guidelines.
   - **Air Quality**: Be aware of ventilation requirements and the use of respirators in areas with poor air quality.
   - **Waste Disposal**: Properly label and dispose of hazardous waste according to regulations to prevent environmental contamination.
2. **Process Hazards**
   - **Standard Operating Procedures (SOPs)**: Follow SOPs to ensure safe and consistent operations. Regularly review and update SOPs as needed.
   - **Process Changes**: Communicate any changes in processes to all affected employees. Ensure everyone understands the new procedures and potential hazards.
   - **Emergency Procedures**: Know the emergency procedures for process-related incidents, including shutdown protocols and evacuation routes.
3. **Equipment Hazards**
   - **Machine Guarding**: Ensure all machinery has appropriate guards in place to prevent accidental contact with moving parts.
   - **LOTO**: Follow LOTO procedures to safely de-energize equipment before maintenance or repair.
   - **Equipment Inspections**: Regularly inspect equipment for wear and tear. Report any defects immediately and do not use faulty equipment.

Group Activity: Hazard Identification and Communication Drill

Objective: To practice identifying hazards and effectively communicating them to the team.

Materials Needed: Hazard identification checklists, sample SDS, PPE examples, and a whiteboard or flip chart.

Instructions:

1. **Divide the group into teams.**
2. **Assign each team a specific area** of the workplace (e.g., storage area, production line, maintenance shop).
3. **Provide hazard identification checklists** to each team.
4. **Have teams inspect their assigned areas** and identify potential environmental, process, and equipment hazards.
5. **Teams should document their findings** and suggest appropriate communication methods (e.g., signage, training, meetings).
6. **Reconvene as a group** and have each team present their findings.
7. **Discuss the identified hazards** and the proposed communication methods. Use the whiteboard or flip chart to compile a comprehensive list of hazards and communication strategies.

Conclusion
Effective hazard communication is vital for preventing accidents and ensuring a safe workplace. By regularly identifying and communicating hazards, we can create a safer environment for everyone. Remember, safety is a shared responsibility.

HEALTHY HABITS

Healthy Habits can help reduce strains and sprains. These injuries are common in the workplace, but with the right lifestyle choices, we can strengthen our bodies and improve our resilience against such injuries.

Benefits of Healthy Habits:

1. Improved Muscle Strength and Flexibility:
   a. Regular exercise helps build strong muscles and improve flexibility, which are crucial for preventing strains and sprains.
   b. Activities like stretching, yoga, and strength training can enhance muscle tone and joint stability.

2. Better Posture and Body Mechanics:
   a. Maintaining good posture and using proper body mechanics during activities can reduce the risk of injury.
   b. Healthy habits such as regular exercise and ergonomic practices at work help support good posture.

3. Enhanced Balance and Coordination:
   a. Exercise that focuses on balance and coordination, such as riding a bike, or balance drills, can help prevent falls and related injuries.
   b. Improved coordination allows for better control of movements, reducing the likelihood of awkward positions that can lead to strains and sprains.

4. Healthy Weight Management:
    a. Maintaining a healthy weight reduces the stress on your joints and muscles, lowering the risk of injury.
    b. A balanced diet and regular physical activity are key components of healthy weight management.

5. Adequate Rest and Recovery:
    a. Ensuring you get enough rest and recovery time allows your muscles and joints to heal and strengthen.
    b. Overworking your body without proper rest can lead to fatigue and increase the risk of strains and sprains.

Healthy Habits to Adopt:

1. Regular Exercise:
    a. Incorporate a mix of cardiovascular, strength, and flexibility exercise.
    b. Aim for at least 30 minutes of focused exercise a day.

2. Proper Nutrition:
    a. Eat a balanced diet rich in fruits and vegetables, lean proteins, and whole grains to support muscle health and recovery.
    b. Stay hydrated by drinking plenty of water throughout the day.

3. Good Posture:
    a. Be mindful of your posture whether sitting, standing, or lifting. Use ergonomic equipment and techniques to support good posture.
    b. Take regular breaks to stretch and move around, especially if you have a sedentary job.

4. Adequate Sleep:
    a. Aim for 7-9 hours of sleep each night to allow your body to recover and repair.
    b. Establish a regular sleep routine and create a restful sleep environment.

5. Warm up- Cool Down:
    a. Always warm up before engaging in activity to prepare your muscles and joints.
    b. Cool down after exercise to help your body recover and reduce muscle stiffness.

Conclusion:
Adopting healthy habits is essential for reducing the risk of strains and sprains. By focusing on regular exercise, proper nutrition, good posture, adequate rest, and proper warm up and cool down routines, we can strengthen our bodies and enhance our overall well-being.

HEART ATTACKS

Heart attacks can happen suddenly and recognizing the symptoms can save lives.

Symptoms:

- **Chest Pain**: Pressure, squeezing, or pain in the center of the chest.
- **Upper Body Discomfort**: Pain in arms, back, neck, jaw, or stomach.
- **Shortness of Breath**: With or without chest discomfort.
- **Other Symptoms**: Cold sweat, nausea, lightheadedness.

Response:

1. **Call 911**: Immediately.
2. **Keep Calm**: Have the person sit and stay calm.
3. **Administer Aspirin**: If not allergic, give aspirin to chew.
4. **Perform CPR**: If unresponsive and not breathing, start CPR.
5. **Use AED**: If available, use as directed.

Conclusion
Quick recognition and response to heart attack symptoms can save lives. Stay informed and ready to act. Stay safe!

HEAT AWARENESS TRAINING

Heat exposure can be a serious risk, especially for those working outdoors or in hot environments. Understanding the dangers and how to prevent heat-related illnesses is crucial for maintaining safety and health.

1. **Types of Heat-Related Illnesses**
   - **Heat Cramps**: Muscle pains or spasms due to heavy sweating.
   - **Heat Exhaustion**: Symptoms include heavy sweating, weakness, dizziness, nausea, and headache.
   - **Heat Stroke**: A severe condition with symptoms like high body temperature, confusion, and loss of consciousness. This is a medical emergency.
2. **Risk Factors**
   - High temperatures and humidity.
   - Direct sun exposure.
   - Physical exertion.
   - Lack of acclimatization.
   - Dehydration.
3. **Prevention Strategies**
   - **Hydration**: Drink plenty of water throughout the day.
   - **Rest Breaks**: Take regular breaks in shaded or cool areas.
   - **Clothing**: Wear light, loose-fitting, and breathable clothing.
   - **Acclimatization**: Gradually increase exposure to hot conditions.
   - **Monitoring**: Keep an eye on coworkers for signs of heat stress.

4. **First Aid for Heat-Related Illnesses**
    - Move the person to a cooler place.
    - Provide water or a sports drink.
    - Apply cool, wet clothes to the skin.
    - Seek medical attention if symptoms are severe.

Personal Heat Exposure Story (Use the Below or Fill in with Your Own)

Let me share a story about a colleague named Alex. Alex was working on a construction site during a particularly hot summer. Despite the high temperatures, Alex was determined to finish his tasks quickly. He skipped a few water breaks and continued working under the scorching sun.

After a few hours, Alex started feeling dizzy and nauseous. His coworkers noticed he was sweating profusely and seemed disoriented. They immediately moved him to a shaded area and gave him water. Recognizing the signs of heat exhaustion, they called for medical help.

Alex was treated for heat exhaustion and learned the importance of taking regular breaks and staying hydrated. His experience served as a valuable lesson for the entire team, emphasizing the need to prioritize health and safety over speed.

When the heat is on, watch out for each other. We all want to make it Safe 2 Home.

HORSE PLAY

Horseplay, or rough and boisterous play, might seem harmless and fun, but it can lead to serious injuries and accidents, especially in retail and warehousing environments. It's important to understand the risks and foster a culture of safety and professionalism.

Key Points:

1. **Understanding Horseplay**:
    - Horseplay includes any form of rough or boisterous behavior that can distract or endanger others.
    - Common examples include running, pushing, throwing objects, and practical jokes.
2. **Risks and Consequences**:
    - **Injuries**: Horseplay can lead to slips, trips, falls, and other injuries. In a warehouse, it can cause accidents involving heavy machinery or falling objects.
    - **Damage to Property**: Rough behavior can result in damaged goods, equipment, and facilities.
    - **Reduced Productivity**: Distractions and accidents caused by horseplay can slow down work processes and reduce overall productivity.
    - **Legal and Financial Implications**: Injuries and damage can lead to legal action and financial losses for both employees and the company.

3. **Promoting a Safe Work Environment**:
   - **Set Clear Expectations**: Establish and communicate clear policies against horseplay. Ensure everyone understands the importance of maintaining a safe work environment.
   - **Lead by Example**: Supervisors and team leaders should model appropriate behavior and discourage horseplay.
   - **Encourage Reporting**: Create an environment where employees feel comfortable reporting unsafe behavior without fear of retaliation.

Story from a Customer

A customer shared their experience of witnessing a horseplay accident in a retail store. While shopping, they saw two employees joking around and chasing each other near the stockroom. One of the employees tripped over a box and fell into a shelf, causing several heavy items to fall. The customer was shocked and concerned about the employee's safety. They felt uneasy about the store's safety practices and questioned the professionalism of the staff. This incident made the customer hesitant to return to the store, highlighting the importance of maintaining a safe and professional environment for both employees and customers.

Questions to Discuss:

1. **What are some examples of horseplay you have witnessed or heard about in our workplace?**
2. **How can horseplay lead to serious accidents or injuries in our environment?**
3. **What steps can we take to discourage horseplay and promote safety?**
4. **How can we support each other in maintaining a safe and professional workplace?**
5. **What should you do if you see someone engaging in horseplay?**

Conclusion

Horseplay has no place in the workplace, especially in environments like retail and warehousing where the risks are high. By understanding the dangers and committing ourselves to a culture of safety, we can protect ourselves and our colleagues from unnecessary harm.

HOUSEKEEPING

Clean as you go, it is something you might have heard us say or talk about, but what does it really mean? **Discuss.**

What Is Housekeeping

It is the maintenance of the working and shopping environment to preserve the safety of those occupying the space. Generally, this maintenance falls into standardized maintenance practices.

Housekeepers are naturally the tip of the spear. They use set routines to keep our environment clean and tidy. From bathrooms to breakrooms. Sweeping, mopping, and vacuuming all our workspace. They also help fight illness, by sanitizing every common touch point, such as doorknobs, counters, and computers. These people have a big job, for which we should all be thankful.

But the dedicated housekeeper can't do this job alone. It takes all of us.

Discuss

How can you help with housekeeping as an individual? As a team, what areas or tasks could we focus on this week/month to make a difference in our work environment?

General Housekeeping Tips:

- Organize parts, products, or other materials as you put them away.
- Do not leave products or materials in walkways. If it is a dedicated walk lane or work area, keep it clear.
- If you do demo work, be sure to remove all nails/screws completely. This prevents snags and punctures.
- All recyclable materials should be sorted and separated from garbage. All waste should be placed in a bin and disposed of properly.
- Damaged equipment should be reported and attended to until Facilities can perform proper LOTO Procedures.
- All hand tools/power tools should be properly stored, or if damaged, removed from service until they are repaired or replaced.
- At the end of your shift, ensure all your equipment, tools, PPE, and personal items are properly stored. If you have any remaining communication needs with your supervisor, contact them directly, or email them with a detailed note. All hazards should be reported before leaving for the day.
- Make sure your work area is ready for the next team member. At any time, your equipment and work area should be ready for a safety inspection.

Discuss

How has this review impacted how you feel about housekeeping at work and at home? Why is Housekeeping as defined important to you?

As we go back to work, remember to Think Safety, and Act Safely.

HYDRATION SAFETY

Staying hydrated is crucial for maintaining health and safety, especially when working in hot environments or performing physically demanding tasks. Proper hydration helps prevent heat-related illnesses and ensures optimal performance.

Importance of Hydration:

- **Regulates Body Temperature**: Water helps regulate your body temperature, especially in hot conditions.

- **Maintains Physical Performance**: Dehydration can lead to fatigue, muscle cramps, and decreased coordination.
- **Supports Cognitive Function**: Staying hydrated helps maintain focus and mental clarity.
- **Prevents Heat-Related Illnesses**: Proper hydration reduces the risk of heat exhaustion and heat stroke.

Signs of Dehydration:

- **Thirst**: Feeling thirsty is an early sign of dehydration.
- **Dry Mouth and Lips**: A dry mouth or chapped lips can indicate a lack of fluids.
- **Dark Urine**: Dark yellow or amber-colored urine is a sign of dehydration.
- **Fatigue**: Feeling unusually tired or lethargic.
- **Dizziness**: Lightheadedness or dizziness can occur with dehydration.

Hydration Tips:

1. **Drink Water Regularly**: Aim to drink water throughout the day, even if you don't feel thirsty.
2. **Carry a Water Bottle**: Keep a water bottle with you and refill it regularly.
3. **Take Breaks**: Take regular breaks to drink water, especially in hot environments.
4. **Eat Hydrating Foods**: Include fruits and vegetables with high water content in your diet, such as watermelon, cucumbers, and oranges.
5. **Avoid Dehydrating Beverages**: Limit intake of caffeinated and sugary drinks, as they can contribute to dehydration.

Conclusion
Staying hydrated is essential for maintaining health and safety on the job. By recognizing the signs of dehydration and following hydration tips, you can prevent heat-related illnesses and ensure optimal performance. Remember to drink water regularly and take care of your body. Stay hydrated and stay safe!

INSECT BITES/STINGS

Insect bites and stings can cause pain, allergic reactions, and even serious medical emergencies. Understanding the dangers, how to avoid bites and stings, and the appropriate medical response is crucial for safety. If possible, take a picture of the insect that caused the bite/sting. This may be beneficial to medical professionals.

Dangers of Insect Bites and Stings:

- **Pain and Swelling**: Most insect bites and stings cause immediate pain, redness, and swelling at the site.
- **Allergic Reactions**: Some individuals may experience severe allergic reactions, including anaphylaxis, which can be life-threatening.

- **Infection**: Scratching the bite or sting site can lead to infection.
- **Multiple Stings**: Multiple stings, especially from bees or wasps, can cause more severe reactions.

How to Avoid Insect Bites and Stings:

1. **Wear Protective Clothing**:
   - Long sleeves, pants, and closed-toe shoes can help protect against bites and stings.
2. **Use Insect Repellent**:
   - Apply insect repellent to exposed skin and clothing.
3. **Avoid Scented Products**:
   - Avoid using scented lotions, perfumes, and hair products that can attract insects.
4. **Be Cautious Around Nests**:
   - Stay away from areas where insects' nest, such as beehives or wasp nests.
5. **Keep Food Covered**:
   - When eating outdoors, keep food and drinks covered to avoid attracting insects.
6. **Stay Calm**:
   - If an insect approaches, it will remain calm and move away slowly. Swatting insects can provoke them.

Medical Response to Insect Bites and Stings:

1. **Remove the Stinger**:
   - If stung by a bee, remove the stinger by scraping it out with a flat object like a credit card. Avoid using tweezers, as squeezing can release more venom.
2. **Clean the Area**:
   - Wash the bite or sting site with soap and water to prevent infection.
3. **Apply Ice**:
   - Apply an ice pack to reduce swelling and pain.
4. **Take Antihistamines**:
   - Over-the-counter antihistamines can help reduce itching and swelling.
5. **Monitor for Allergic Reactions**:
   - Watch for signs of an allergic reaction, such as difficulty breathing, swelling of the face or throat, dizziness, or a rapid heartbeat.
6. **Seek Medical Attention**:
   - If the person shows signs of a severe allergic reaction (anaphylaxis), call 911 immediately. Use an epinephrine auto-injector if available.

Conclusion

Insect bites and stings can be more than just a nuisance; they can pose serious health risks. By taking preventive measures and knowing how to respond to bites and stings, we can protect ourselves and others. Stay vigilant and stay safe!

INCIDENT REPORTING

Employees will immediately inform their supervisor of any incident which injures, damages, or recklessly endangers people or property. Employees who decline to report such incidents are subject to disciplinary action, possible denial or reduction of worker's compensation benefits and any other claims.

Company policy states the requirement to report all incidents including near misses or reckless endangerment. Incident reporting is the key tool in preventing the same incident from happening again.

Types of Incidents:

- Employee Injury
- Employee Property damage
- Company Property damage
- Public involved incident
- Third Party Damage
- Theft
- Near Miss/ Reckless Endangerment

Reporting Procedure:

1. Employee reports to their supervisor or the nearest supervisor on site.
2. Employee contacts Authorities if necessary, and/or the company Security/ Safety officers.
3. Any witnesses will be asked to complete an Incident Witness Report or IWR.
4. Supervisors will complete the Incident Report and First Report of Injury Report within 1 hour of Fatal or amputation incidents, and with 4 hours of an incident causing hospitalization. The remaining incidents should be reported within 24 hours.

Hazard Conditions

All unsafe or dangerous conditions, damaged equipment, or reckless behavior should be reported immediately to a supervisor. If the situation cannot be remedied; the area, tools, or equipment will be isolated/closed until conditions can be made safe.

JOB HAZARD ANALYSIS

A Job Hazard Analysis (JHA) is a technique that focuses on job tasks to identify hazards before they occur. It involves breaking down a job into its individual steps and analyzing each step for potential hazards.

1. **Steps to Conduct a JHA**:
   - **Select the Job**: Choose a job that has a history of incidents or one that is particularly hazardous.
   - **Break Down the Job**: List all the steps involved in the job, from start to finish.

- **Identify Hazards**: Examine each step to identify potential hazards. Consider what could go wrong, the consequences, and how likely it is to happen.
- **Develop Solutions**: Determine ways to eliminate or control hazards. This could include engineering controls, administrative controls, or personal protective equipment (PPE).

2. **Employee Involvement**:
   - Employees should actively participate in the JHA process. They have firsthand knowledge of the tasks and can provide valuable insights into potential hazards and effective controls.
3. **Reporting Concerns**:
   - If you identify any hazards or have concerns about safety while performing your tasks, report them immediately to your supervisor or safety officer. Prompt reporting can prevent accidents and injuries.

How to Examine Your Tasks/Job:

- **Observe and Document**: Watch yourself or a coworker perform the job and note each step.
- **Ask Questions**: Consider what could go wrong at each step. What are the potential hazards? What are the consequences if something goes wrong?
- **Evaluate Controls**: Think about existing controls and whether they are sufficient. Are there additional measures that could be implemented to enhance safety?

Discussion Points:

- **What are some common hazards you encounter in your daily tasks?**
- **How can we improve our current hazard controls to ensure a safer work environment?**

KNIFE SAFETY

Utility Knives, hook knives, and switch blade knives are the essential tool used by more than 80% of workers worldwide. No wonder why they account for **40% of all workplace tool injuries!**

Consider providing safety enhanced box cutters with Auto Retractable blades. This specific tool was chosen because it effectively removes two mechanical dangers standard box knives and switch blades have.

Do You Know What Those Two Dangers Are:

- The auto retrack button requires the use of your thumb. This forces the user to have four fingers around the tool securing it, and keeps the thumb out of the cutting path

- The auto retrack button pulls the blade back into the tool if the user loses grip on the tool or stops using it. 10% of blade cuts happen when the victim was unaware the knife was present.

Knife How To: Can You Name Three Knife Safety Tips:

- Always use a sharp blade
- Always cut horizontally, not vertically
- Always cut away from yourself
- Avoid distractions when using a knife
- Always store your knife when not in use
- Do not force the blade through more material than it can safely cut
- Ensure the knife blade will not and does not impact a surface or object that should not be cut

Cutting Plastic Bags:

- Pull the bag away from the product with non-cutting hand
- Insert blade in bag, 6 inches from your non cutting hand, facing away from your hand.
- Gently slide the blade through the material
- Occasionally changing the position of the non-cutting hand to keep the material away from the product being removed from the bag
  - A hook knife is a safer alternative to using a box cutter for this. The same cutting technique should be utilized.

Knife Protection

Proper cutting technique should always be utilized when using a box cutter or hook knife. Following the tips provided in this tool box talk are best practice. Additional PPE is also available and provided upon request. Cut resistant gloves and eye wear are considered standard PPE when opening products with a box cutter or hook knife.

A knife should be used for cutting only. Never use a knife as a screw driver or prying tool.

Use the RIGHT TOOL

Using the right type of knife in the proper way is essential to our safety. This same ideology can be applied to any situation where a tool would give us an advantage on the job, in either safety, or production. Never attempt to open a box or bag without the correct tools. Opening a taped box by hand without cutting the tape, could create unnecessary strain on your fingers, elbow, and shoulder joints. Using the correct tool for each job will ensure you leave here every day ready to take on the most important jobs at home.

Watch out for each other and let's all make it home safe.

LEAVE YOURSELF AN OUT

Introduction

Injuries at work often occur when individuals put themselves in harm's way or find themselves unable to escape danger. Many incidents, such as a dropped load falling on your foot or rear-ending the car in front of you, are preventable if you leave yourself an out.

Defining "Leaving Yourself an Out"

"Leaving yourself an out" means having the ability to escape danger if a negative situation occurs. This concept is frequently emphasized in defensive driving courses, which stress the importance of always being prepared to react to changing conditions and avoiding preventable collisions.

Examples of "Leaving Yourself an Out" at Work

- **Dropped Loads**: Never position yourself where a falling load can strike you. Always give extra space for unexpected changes or failures.
- **Working Around Vehicles or Equipment**: Ensure operators of heavy equipment or vehicles can see you. If you have an out, you can escape danger even if you find yourself in the line of fire.

Summary

Always avoid putting yourself directly in the line of fire, such as under a lifted load or behind a backing vehicle. Consider your options if conditions change and a negative situation occurs. Never put yourself in a position where you cannot escape danger. Always leave yourself an out.

Discussion Points:

1. **Why is it important to always leave yourself an out in various work scenarios?**
2. **Can you share an example of a situation where leaving yourself an out prevented an injury?**
3. **What are some strategies to ensure you always have an out in your specific work area?**

LET IT GO

In the workplace, conflicts and frustrations are inevitable. However, holding onto grudges or dwelling on minor issues can create a negative environment and impact productivity. Developing the maturity to "let it go" is essential for maintaining a positive and collaborative workplace.

Importance of Letting Go:

- **Reduces Stress**: Holding onto negative emotions can increase stress and affect your mental health.

- **Improve Relationships**: Letting go of minor conflicts helps build stronger, more cooperative relationships with colleagues.
- **Enhance Focus**: By not dwelling on past issues, you can focus better on your tasks and goals.
- **Promotes a Positive Environment**: A workplace where employees can move past conflicts is more harmonious and productive.

Strategies for Letting Go:

1. **Perspective**:
   - Consider the bigger picture. Ask yourself if the issue will matter in the long run.
2. **Communication**:
   - Address conflicts directly and calmly. Clear communication can resolve misunderstandings and prevent resentment.
3. **Empathy**:
   - Try to understand the other person's point of view. Empathy can help you see the situation differently and reduce negative feelings.
4. **Focus on Solutions**:
   - Instead of dwelling on the problem, focus on finding a solution. This proactive approach can help you move forward.
5. **Self-Care**:
   - Engage in activities that help you relax and de-stress, such as exercise, hobbies, or spending time with loved ones.

Discussion Points:

1. **Personal Experiences**:
   - Can you share a time when you held onto a minor issue at work? How did it affect you and your work environment?
2. **Strategies**:
   - What strategies have you found effective in letting go of conflicts or frustrations? How can we support each other in this effort?
3. **Benefits**:
   - How can letting go of minor issues improve our workplace culture and productivity?

Conclusion

Developing maturity to "let it go" is crucial for a positive and productive work environment. By focusing on perspective, communication, empathy, solutions, and self-care, we can move past conflicts and work more harmoniously. Let's commit to fostering a supportive and collaborative workplace. Stay positive and stay productive!

LOCK OUT TAG OUT

LOTO procedures may seem like they do not affect you, but let's take a look at why it's one of the most important safety topics we can cover.

Have you ever been working at home, maybe in the garage or basement and someone accidentally shut the lights off in the space? Suddenly you are in total darkness, and your immediate reaction is to… . The common answer is shout, hey turn the lights on. Sure, this time no harm was done, but what if you were working with a saw or other piece of equipment? In that situation, losing power unexpectedly can be dangerous, but power coming on unexpectedly can be deadly.

These types of incidents happen all too often and have resulted in the need for a LOTO program. Shutting off the machine simply is not enough protection.

What Basic Energized Equipment Hazards Are Out There?
Mechanical, Thermal, Pneumatic, Hydraulic, Fluids/Gases, Gravity, Electrical, and Water under pressure.

An example of a mechanical danger is the pulley gear on an order picker mast.
An example of Hydraulic Danger is the garbage compactors.

Most machines offer multiple hazards or dangers during operation. This further requires the LOTO process to ensure safety.

The Six Basic Steps for LOTO:

1. Prepare for shut down, know the type of energy being isolated, its hazards, and how to control it.
2. Follow normal procedures for machine and equipment shut down.
3. Turn off the energy at the main power source.
4. Lock Out or Tag Out all energy sources.
5. Release stored energy within the system.
6. Verify the isolation by attempting to cycle/start the machine, then return operating controls to their off position.

Four Basic Steps for Releasing Equipment from LOTO:

1. Inspect machine or equipment for non-essential items/replace guards.
2. Conduct a head count and safely position employees away from the area.
3. Remove the LOTO device.
4. Notify affected personnel that the LOTO device was removed.

It is then the responsibility of the equipment's operator to safely test the equipment before returning to normal production activities (29 CFR 1910.147, The Control of Hazardous Energy (Lockout/Tagout, 2025).

Who Should Lock Out and Tag Out Equipment
The Facilities team is trained in proper LOTO procedures. If a piece of equipment needs to be shut down and LOTO put into effect, Facilities must be contacted. If you are unable to contact them, let your supervisor know immediately and do not leave the malfunctioning equipment until you are given permission to do so by a supervisor or Facilities team member.

Tampering with or removing a LOTO device without permission from the individual who did the LOTO is never allowed. This is poor practice that could result in a dangerous piece of equipment being used by someone without knowledge of the danger.

At this point, discuss as a team at least three instances where they may find the need to request a device undergo the LOTO procedures.

Remember we all want to make it **Safe 2 Home**. Please always follow LOTO procedures.

MACHINE GUARDING

Machine guards are essential for protecting workers from the dangerous moving parts of machinery. Proper use and maintenance of these guards can prevent serious injuries such as cuts, amputations, and crush injuries (29 CFR 1910.212 General Requirements of all Machines, 2025).

Discussion Points:

1. **Types of Machine Guards**
   - **Fixed Guards**: Permanently attached to the machine and provide a barrier between the operator and the moving parts.
   - **Interlocking Guards**: Automatically shut off or disengage the machine when the guard is opened or removed.
   - **Adjustable Guards**: Can be adjusted to fit different sizes of material being worked on.
   - **Self-Adjusting Guards**: Move according to the size and movement of the material.
2. **Purpose of Machine Guards**
   - **Prevent Contact**: Reduce the possibility of the operator or a co-worker coming into contact with moving parts.
   - **Protect from Falling Objects**: Prevent objects from falling into the machine and causing injuries.
   - **Enhance Safety**: Ensure the machine operates safely and efficiently.
3. **Inspection and Maintenance**
   - **Regular Inspections**: Check the guards regularly to ensure they are in good condition and properly secured.
   - **Report Issues**: Immediately report any damaged or missing guards to a supervisor.
   - **Maintenance**: Follow the manufacturer's guidelines for maintaining and replacing guards.
4. **Safe Operation Practices**
   - **Never Remove Guards**: Do not remove or bypass guards while the machine is in operation.
   - **Proper Use**: Use the machine only for its intended purpose and follow all safety instructions.

- **Training**: Ensure all operators are trained in the proper use and importance of machine guards.

5. **Personal Protective Equipment (PPE)**
   - **Wear Appropriate PPE**: Use gloves, safety glasses, and other PPE as required to protect against hazards.

Group Activity: Machine Guard Inspection Drill:

Objective: To practice identifying and inspecting machine guards.
Materials Needed: Inspection checklists, sample machine guards, and PPE examples.

Instructions:

1. **Divide the group into teams**.
2. **Assign each team a specific machine** or area with machinery.
3. **Provide inspection checklists** for each team.
4. **Have teams inspect their assigned machines** for proper guard installation and condition.
5. **Teams should document their findings** and report any issues.
6. **Reconvene as a group** and discuss the findings.
7. **Review the importance of each type of guard** and proper maintenance practices.

Conclusion
Machine guards are a critical component of workplace safety. Regular inspections, proper maintenance, and adherence to safety protocols can prevent accidents and injuries. Remember, safety is everyone's responsibility.

MATERIAL HANDLING

Lifting and Transporting materials are the most common cause of strain and sprain injuries. Improper handling increases the risk of injury. Additional information can be found in the various Ergonomic training courses offered in this text and online.

Lifting:

- Loads exceeding 65lbs, loads that are large, or awkward often require lifting aids, or two or more people.
- When lifting we want to use the Attack or Sumo Stance. Hinge at the knees, back straight, head in line with the spine. Feet slightly more than shoulder width apart.
- Get a grip. Up to 80 of your lifting capacity is dependent on your grip. Rubber palmed gloves can be a huge advantage.
- Avoid the twist. Twisting while lifting or moving loads the joints and can potentially damage tendons and cartilage.

- Keep it tight. The load should be kept tight to your body. A secure load prevents reactionary movements which can cause injuries.
- Never lift over your chest height. Extending the load above your center of gravity minimizes your stability and can cause injury.
- Avoid moving loads from the ground. When possible, store heavy items at or above knee height and below shoulder height.

Transporting:

- Vision is essential to safely transporting materials over any distance. Vision helps ensure the pathway is clear and free of hazards.
- In some situations, it may be necessary to clear the route or path before picking up and relocating the item.
- If possible, utilize mechanical tools such as a two-wheel cart or other transportation aid when moving materials more than 10-15 feet.

Physical Material Handling requires close attention to the environment and to the material being handled. Watching out for others and coaching them through proper lifting and moving techniques is essential to creating a safe environment for all.

Discuss

How can we help others by identifying dangerous lifts and material movements? What processes can we change or improve in order to reduce exposure to risk?

MEZZANINE SAFETY

Mezzanines are elevated platforms often used in warehouses and industrial settings to maximize storage space. While they are useful, mezzanines also pose significant safety risks if not properly managed. Understanding these dangers and following safety protocols can prevent accidents and injuries.

Key Points:

1. **Fall Hazards**:
   - **Edge Protection**: Ensure mezzanines have guardrails, handrails, and toe boards to prevent falls.
   - **Openings**: Cover or guard any openings, such as stairwells or access points, to prevent accidental falls.
   - **Personal Fall Protection**: Use personal fall protection equipment, such as harnesses and lanyards, when working near the edge or on elevated platforms.
2. **Structural Integrity**:
   - **Load Limits**: Adhere to the load limits specified by the manufacturer. Overloading can compromise the structural integrity of the mezzanine.

- **Regular Inspections**: Conduct regular inspections to check for signs of wear, damage, or corrosion. Report any issues immediately.
- **Professional Installation**: Ensure mezzanines are installed and modified by qualified professionals according to the manufacturer's guidelines.

3. **Access and Egress**:
   - **Safe Access**: Use designated stairs or ladders to access mezzanines. Ensure they are in good condition and free of obstructions.
   - **Clear Pathways**: Keep pathways on mezzanines clear of clutter and obstacles to prevent trips and falls.
   - **Emergency Exits**: Ensure there are clear, and accessible emergency exits from mezzanines.
4. **Material Handling**:
   - **Safe Lifting**: Use appropriate lifting techniques and equipment to move materials to and from mezzanines.
   - **Secure Loads**: Ensure materials are securely stored and do not overhang the edges of the mezzanine.
   - **Mechanical Aids**: Use mechanical aids, such as forklifts or hoists, to lift heavy items to mezzanines safely.
5. **Training and Awareness**:
   - **Employee Training**: Provide training on mezzanine safety protocols and the proper use of personal fall protection equipment.
   - **Signage**: Post clear signage indicating load limits, access points, and safety requirements.
   - **Safety Culture**: Foster a culture of safety where employees feel comfortable reporting hazards and unsafe conditions.

Conclusion

Mezzanines can be valuable additions to the workplace, but they come with inherent risks. By understanding the dangers and following safety protocols, we can prevent accidents and ensure a safe working environment. Always adhere to load limits, use proper fall protection, and keep pathways clear. Let's commit to maintaining mezzanine safety and protecting ourselves and our colleagues.

MOBILE STEPS

Mobile Steps have a great deal of value when being used to assist with loading and unloading tasks. Mobile steps enhance safety and efficiency when accessing elevated areas or handling heavy items.

Benefits of Using Mobile Steps:

1. Improved Safety:
   a. Mobile steps provide a stable and secure platform for reaching higher areas, reduce the risk of falls, and eliminate potential fall damage to product/equipment.

   b. Mobile steps allow the user to lift and move items at chest height, eliminating nearly all overhead lifting.
   c. They may be equipped with handrails and non-slip treads, offering additional support and stability.
2. Enhanced Efficiency:
   a. Using mobile steps allows you to access elevated areas quickly and easily, speeding up the loading and unloading process.
   b. They can be moved to different locations as needed, providing flexibility and convenience.
3. Reduced Physical Strain:
   a. Mobile steps minimize the need to stretch or overreach objects, which can lead to muscle strain and fatigue.
   b. They allow you to maintain proper posture while working, reducing the risk of back, shoulder, and neck injuries.

How to Use Mobile Steps Safely:

1. Inspect the Steps:
   a. Before using it, check the steps for any signs of damage or wear. Ensure the steps are stable and wheels locked in place.
   b. Make sure the handrails are secure, and the treads clean of debris.
2. Position the steps correctly:
   a. Place the steps on a flat, stable surface. Ensure that the steps are positioned close to the area needing access.
   b. Reposition frequently. Keeping the position relative to the task.
   c. Always work on steps facing the direction of travelling upstairs. Never work off the side of a set of stairs.
   d. Avoid carrying heavy or bulky items while traversing the stairs. Ask for assistance.
3. Be aware of your surroundings:
   a. Ensure the area around the steps/stairs is clear of obstacles and hazards.
   b. Communicate with your team to ensure everyone is aware of your movements and the location of the steps.

Conclusion

Using mobile steps can significantly improve the safety of your operations during loading and unloading tasks. Steps also improve the efficiency of tasks and improve workflow. By following proper techniques and being mindful of your surroundings, we can reduce the risk of injuries and enhance productivity. Please use those mobile steps and take advantage of this benefit.

NOT MY PROBLEM

The "Not My Problem" mindset can be detrimental to workplace safety and productivity. This attitude can lead to unresolved issues, increased risks, and a negative work environment. It's important to understand why this mindset is harmful and how to foster a culture of responsibility and teamwork.

Why the "Not My Problem" Mindset Is Harmful

- **Safety Risks**: Ignoring potential hazards can lead to accidents and injuries.
- **Decreased Productivity**: Problems left unaddressed can disrupt workflow and efficiency.
- **Negative Work Environment**: This mindset can create a culture of apathy and disengagement.
- **Lack of Accountability**: Failing to take responsibility can lead to a breakdown in trust and cooperation among team members.

Promoting a Culture of Responsibility

1. **Encourage Ownership**:
   - Take responsibility for identifying and addressing issues, even if they fall outside your immediate duties.
2. **Foster Teamwork**:
   - Work collaboratively to solve problems and support each other in maintaining a safe and productive environment.
3. **Communicate Openly**:
   - Share concerns and suggestions with supervisors and colleagues to ensure issues are addressed promptly.
4. **Lead by Example**:
   - Demonstrate a proactive attitude by taking the initiative and showing that you care about the well-being of the team and the workplace.

Discussion Prompts

1. **Identifying Issues**:
   - Can you think of a time when the "Not My Problem" mindset led to a safety issue or disrupted work? How could it have been handled differently?
2. **Taking Initiative**:
   - What are some ways you can take the initiative to address problems or hazards in our workplace? How can we support each other in this effort?

Conclusion

The "Not My Problem" mindset can have serious consequences for workplace safety and productivity. By promoting a culture of responsibility, teamwork, and open communication, we can create a safer and more positive work environment. Let's commit to taking ownership and supporting each other in addressing issues. Stay proactive and stay safe!

OFFICE SAFETY

Office equipment, such as computers, printers, copiers, and shredders, are essential for daily operations. However, improper use or maintenance of this equipment can lead to accidents and injuries. Understanding and following safety protocols is crucial for maintaining a safe office environment.

Key Points:

1. **Ergonomics**:
    - **Proper Setup**: Ensure your workstation is set up ergonomically. Adjust your chair, monitor, and keyboard to reduce strain on your body.
    - **Posture**: Maintain good posture while sitting. Keep your feet flat on the floor, and your back supported.
    - **Breaks**: Take regular breaks to stretch and move around to prevent repetitive strain injuries.
2. **Electrical Safety**:
    - **Power Cords**: Keep power cords organized and out of walkways to prevent tripping hazards. Avoid overloading power strips and outlets.
    - **Equipment Maintenance**: Regularly inspect office equipment for frayed cords, exposed wires, or other damage. Report any issues immediately.
    - **Unplugging**: Always unplug equipment before cleaning or servicing it.
3. **Proper Use of Equipment**:
    - **Printers and Copiers**: Follow the manufacturer's instructions for loading paper and replacing toner or ink cartridges. Avoid placing hands near moving parts.
    - **Shredders**: Use shredders according to the manufacturer's guidelines. Keep hands, clothing, and jewelry away from the opening of the feed.
    - **Computers**: Avoid eating or drinking near computers to prevent spills. Keep liquids away from all electronic equipment.
4. **Fire Safety**:
    - **Fire Extinguishers**: Know the location of fire extinguishers and how to use them. Ensure they are regularly inspected and maintained.
    - **Emergency Exits**: Keep emergency exits clear of obstructions. Know the evacuation plan and participate in regular fire drills.
    - **Smoke Detectors**: Ensure smoke detectors are installed and function properly.
5. **General Safety Tips**:
    - **Storage**: Store heavy items on lower shelves to prevent them from falling. Use step stools or ladders to reach high shelves safely.
    - **Cleanliness**: Keep your workspace clean and free of clutter to reduce the risk of accidents.
    - **Training**: Participate in training sessions on the safe use of office equipment and emergency procedures.

Activity: Office Safety Scavenger Hunt
To reinforce the importance of office equipment safety, we will conduct a scavenger hunt. This activity will help everyone become familiar with safety features and protocols in our office.

Instructions:

1. **Form Teams**: Divide into small teams.
2. **List of Items**: Each team will receive a list of safety-related items and locations to find, such as:
   - Fire extinguishers
   - Emergency exits.
   - First aid kits
   - Ergonomic workstation setups
   - Power strips and cord management solutions
   - Properly stored heavy items
3. **Time Limit**: Set a time limit for the scavenger hunt (e.g., 20 minutes).
4. **Completion**: Teams must take a photo or note the location of each item on the list.
5. **Debrief**: After the scavenger hunt, gather to discuss what was found and reinforce the importance of each safety feature.

Conclusion
Office equipment safety is essential for preventing accidents and ensuring a productive work environment. By following ergonomic guidelines, practicing electrical safety, using equipment properly, and being prepared for emergencies, we can maintain a safe office. Let's commit to these safety practices and look out for each other.

OFFICE ERGONOMICS

Good morning, everyone! Today, we are going to talk about office ergonomics. Ergonomics is all about designing our work environment to fit us better, which can help prevent discomfort and injuries. By making a few adjustments to our workstations and habits, we can improve our comfort and productivity.

Discussion Points:

1. **Chair and Posture**
   - **Adjustable Chair**: Ensure your chair is adjustable and provides good lumbar support. Your feet should be flat on the floor or on a footrest.
   - **Posture**: Sit back in your chair with your back straight and shoulders relaxed. Avoid slouching or leaning forward for extended periods.
2. **Desk and Monitor Setup**
   - **Monitor Height**: Position your monitor so the top of the screen is at or slightly below eye level. This helps reduce neck strain.
   - **Distance**: Keep the monitor about an arm's length away from your eyes.
   - **Desk Height**: Your desk should be at a height where your elbows are at a 90-degree angle when typing.
3. **Keyboard and Mouse Placement**
   - **Keyboard Position**: Place your keyboard directly in front of you. Your wrists should be straight and your hands at or slightly below elbow level.

- **Mouse Placement**: Keep your mouse close to your keyboard to avoid overreaching. Use a mouse pad with wrist support if needed.

4. **Breaks and Movement**
   - **Regular Breaks**: Take short breaks every hour to stand up, stretch, and move around. This helps reduce muscle fatigue and improve circulation.
   - **Stretching Exercises**: Incorporate simple stretching exercises into your routine to relieve tension in your neck, shoulders, and back.
5. **Lighting and Screen Glare**
   - **Proper Lighting**: Ensure your workspace is well-lit to reduce eye strain. Use task lighting if necessary.
   - **Reduce Glare**: Position your monitor to avoid glare from windows or overhead lights. Use an anti-glare screen if needed.
6. **Personal Comfort**
   - **Temperature and Ventilation**: Adjust the temperature and ventilation to your comfort level. Use a fan or heater if necessary.
   - **Hydration**: Keep a water bottle at your desk and stay hydrated throughout the day.

Activity: Ergonomic Assessment
Let us take a few minutes to assess our own workstations. Follow these steps:

1. **Chair Check**: Adjust your chair height so your feet are flat on the floor and your knees are at a 90-degree angle. Ensure your back is supported.
2. **Monitor Check**: Position your monitor so the top of the screen is at or slightly below eye level. Adjust the distance to about an arm's length away.
3. **Keyboard and Mouse Check**: Ensure your keyboard and mouse are positioned so your wrists are straight, and your hands are at or slightly below elbow level.
4. **Lighting Check**: Adjust your monitor or lighting to reduce glare. Ensure your workspace is well lit.
5. **Posture Check**: Sit back in your chair with your back straight and shoulders relaxed. Make any necessary adjustments to maintain good posture.

Take a moment to make these adjustments. If you need any help or have questions, feel free to ask!

Conclusion
By paying attention to office ergonomics, we can create a more comfortable and productive work environment. Let us all take a moment to assess our workstations and make any necessary adjustments. If you have any questions or need help with your setup, feel free to ask. Let us work together to stay healthy and comfortable!

OPENING AND EDGE FALL HAZARDS

Falls are one of the leading causes of workplace injuries and fatalities. Openings and edges present significant fall hazards that require proper safety measures. This talk will cover key points to ensure safety around these areas.

1. **Identify Fall Hazards**
   - **Openings**: Floor holes, skylights, and roof openings are common fall hazards.
   - **Edges**: Unprotected edges of roofs, platforms, and scaffolds pose serious risks.
2. **Use Guardrails and Covers and Fall Netting**
   - **Guardrails**: Install guardrails around openings and edges to prevent falls.
   - **Covers**: Use sturdy covers over floor openings. Ensure covers are securely fastened and labeled.
   - **Fall Netting**: Rated Netting used at or just below fall edges as designated by OSHA.
3. **Personal Fall Arrest Systems (PFAS)**
   - **Edge Protection:** Traditional Edge protection is preferred. PFAS should be used when Edge Protection is not an option.
   - **Harnesses and Lanyards**: Use appropriate PFAS when working near openings and edges.
   - **Anchor Points**: Ensure anchor points are secure and capable of supporting the load.
4. **Safe Work Practices**
   - **Stay Alert**: Always be aware of your surroundings and the location of openings and edges.
   - **Communication**: Communicate with coworkers about potential fall hazards and safety measures.
5. **Training and Awareness**
   - **Regular Training**: Provide regular training on fall protection and hazard recognition.
   - **Emergency Procedures**: Ensure everyone knows the procedures for responding to fall incidents.
6. **Inspection and Maintenance**
   - **Regular Inspections**: Inspect guardrails, covers, and PFAS regularly for damage or wear.
   - **Maintenance**: Repair or replace damaged safety equipment immediately.

Discussion Points:

1. **What are some common fall hazards you encounter in your work area?**
2. **How do you ensure guardrails and covers are properly installed and maintained?**
3. **Can you describe the steps you take to use a personal fall arrest system correctly?**
4. **What should you do if you notice a missing or damaged guardrail or cover?**
5. **How do you communicate potential fall hazards to your coworkers?**

Conclusion

By following these safety practices and discussing potential hazards, we can reduce the risk of falls and create a safer working environment. Always stay vigilant and report any safety concerns to your supervisor.

ORDER PICKER SAFETY

Good morning, everyone! Today, we are discussing order picker safety. These machines are essential but can be dangerous if not used properly. Let us go over some key safety points.

Discussion Points:

1. **Pre-Operational Checks**
    - **Visual Inspection**: Check for damage or wear.
    - **Fluid Levels**: Ensure battery water is at the correct level.
    - **Battery Check**: Make sure the battery is charged and connected.
2. **Safe Operation**
    - **Harness Use**: Always wear a safety harness.
    - **Speed Control**: Operate at a safe speed.
    - **Load Handling**: Ensure loads are stable and secure.
    - **Visibility**: Maintain a clear view of your path.
3. **Workplace Awareness**
    - **Pedestrian Safety**: Be aware of pedestrians and use the horn to alert them.
    - **Aislc Clcarancc**: Kccp aislcs clcar of obstructions.
    - **Floor Conditions**: Avoid wet or slippery areas.
4. **Loading and Unloading**
    - **Proper Stacking**: Stack loads evenly and securely.
    - **Fork Positioning**: Position forks properly under the load.
    - **Loading Docks**: Ensure dock plates are secure.
5. **Parking and Shutdown**
    - **Safe Parking**: Park in designated areas, lower forks, set the brake, and turn off the engine.
    - **Key Removal**: Remove the key when not in use.

Inspection Activity
Let us perform a quick inspection:

1. **Visual Inspection**: Check for damage.
2. **Battery Check**: Ensure the battery is charged.
3. **Operational Test**: Test brakes, steering, and lift functions.
4. **Safety Features**: Test horn and lights.

Conclusion
Order picker safety is crucial. Follow these guidelines and conduct regular inspections to stay safe. If you have any questions or need further training, please ask. Safety is everyone's responsibility!

OVERHEAD POWER LINES AND DELIVERY TRUCKS

Overhead power lines pose significant hazards for delivery trucks, especially when loading and unloading materials. It's crucial to understand the risks and follow safety procedures to prevent accidents and injuries. Let us cover a few key points about overhead power line safety.

Importance of Overhead Power Line Safety

Overhead power lines carry high voltage electricity, which can cause severe injuries or fatalities if contacted. Delivery truck drivers must be aware of their surroundings and take precautions to avoid contact with power lines.

Key Safety Precautions

1. **Awareness and Planning**
   - **Identify Power Lines**: Before starting any delivery, identify the location of overhead power lines.
   - **Plan Routes**: Plan your route and delivery operations to avoid areas with low-hanging power lines.
   - **Weather**: Weather such as high winds, and ice can impact power lines and cause them to sag or contact points lower than normal.
2. **Safe Distance**
   - **Maintain Clearance**: Always maintain a safe distance from power lines. The recommended minimum clearance is 10 feet.
   - **Use a Spotter**: When working near power lines, use a spotter to help guide you and ensure safe distances are maintained.

****Pay particular attention to Farmyards, private business lots, and apartment complexes**. **

3. **Equipment and Vehicle Safety**
   - **Lower Equipment**: Ensure that all equipment is fully lowered before moving under power lines.
   - **Check Height**: Know the height of your vehicle and any equipment to ensure it can pass safely under power lines.
4. **Emergency Procedures**
   - **Stay Inside**: If your vehicle contacts a power line, stay inside the vehicle and call for help. Do not exit the vehicle until the power company confirms it is safe.
   - **Warn Others**: Warn others to stay away from the vehicle and the power line until the situation is resolved.

Storage Recommendations

- **Designated Areas**: Store delivery trucks and equipment in designated areas away from overhead power lines.
- **Regular Inspections**: Conduct regular inspections of storage areas to ensure they remain clear of power line hazards.

Conclusion

Understanding the hazards associated with overhead power lines and following proper safety practices can prevent accidents and injuries. Always stay vigilant and report any safety concerns to your supervisor.

Quiz

1. What is the recommended minimum clearance distance from overhead power lines?
   - a) 5 feet
   - b) 10 feet
   - c) 15 feet
2. What should you do if your vehicle contacts a power line?
   - a) Exit the vehicle immediately
   - b) Stay inside the vehicle and call for help
   - c) Try to move the vehicle away from the power line
3. Why is it important to use a spotter when working near power lines?
   - a) To help guide you and ensure safe distances are maintained
   - b) To speed up the delivery process
   - c) To assist with loading and unloading materials
4. What should you do before moving a delivery truck under power lines?
   - a) Ensure all equipment is fully lowered
   - b) Increase the speed of the vehicle
   - c) Sound the horn to alert others

Your safety depends on your awareness of the dangers we take for granted every day. Stay Vigilant.

OWNING SAFETY

Owning safety means taking personal responsibility for your own safety and the safety of those around you. It's about being proactive, vigilant, and committed to creating a safe work environment. Let's discuss the importance of owning safety and share a story to illustrate its impact.

Importance of Owning Safety

- **Prevents Accidents**: Taking responsibility for safety helps prevent accidents and injuries.
- **Creates a Safety Culture**: When everyone owns safety, it fosters a culture where safety is a priority.
- **Empowers Employees**: Owning safety empowers employees to speak up and take action when they see potential hazards.
- **Improves Compliance**: Personal responsibility ensures that safety protocols and regulations are followed consistently.

Story: The Importance of Owning Safety

A few years ago, a manager with another company shared the story of working on the floor one day, when he noticed that a stack of heavy boxes was leaning precariously. Instead of assuming someone else would take care of it, he decided to report it immediately. The warehouse supervisor inspected the stack and found that it was indeed unstable. The area was cordoned off, and the boxes were restacked properly.

A few days later, he learned that another warehouse employee had experienced a similar situation where a stack of boxes fell, resulting in serious injuries. It was a stark reminder of how important it is to own safety. By taking action and reporting the hazard, we potentially prevented a similar incident in our warehouse. This experience reinforced my belief that everyone has a role to play in maintaining a safe work environment.

How to Own Safety

1. **Be Proactive**:
    - Identify and address potential hazards before they become problems.
2. **Follow Protocols**:
    - Adhere to all safety procedures and guidelines consistently.
3. **Speak Up**:
    - Report unsafe conditions or behaviors immediately.
4. **Stay Informed**:
    - Keep up to date with safety training and best practices.
5. **Encourage Others**:
    - Promote a culture of safety by encouraging your colleagues to take responsibility for their own safety.

Conclusion

Owning safety is crucial for preventing accidents and creating a safe work environment. By being proactive, following protocols, speaking up, staying informed, and encouraging others, we can all contribute to a culture of safety. Let's commit to owning safety every day. Stay vigilant and stay safe!

PALLET JACK SAFETY

Pallet jacks are essential tools for moving heavy loads in warehouses and other settings. However, improper use can lead to injuries. This talk will cover the basics of safe pallet jack operation.

Discussion Points

1. **Personal Protective Equipment (PPE)**
    - **Wear gloves** to protect your hands from splinters and sharp edges.
    - **Safety glasses** can prevent eye injuries from debris.
2. **Pre-Use Inspection**
    - **Check the wheels** for damage or debris.

- **Inspect the forks** for cracks or bends.
- **Test the controls** to ensure they are functioning properly.

3. **Loading the Pallet Jack**
   - **Do not exceed the load capacity** of the pallet jack.
   - **Center the load** on the forks to maintain balance.
   - **Lift slowly** to avoid sudden movements that could cause the load to shift.
4. **Operating the Pallet Jack**
   - **Push don't pull** the pallet jack to reduce strain on your back.
   - **Move at a safe speed** to maintain control.
   - **Keep your feet clear** of the pallet jack to avoid crushing injuries.
5. **Navigating the Workplace**
   - **Plan your route** to ensure it is free of obstacles.
   - **Use caution on inclines**: move in reverse when going downhill.
   - **Communicate with coworkers** to avoid collisions.
6. **Storing the Pallet Jack**
   - **Store in a designated area** to prevent tripping hazards.
   - **Lower the forks** to the ground when not in use.

Conclusion

By following these safety guidelines, we can minimize the risk of injury and ensure efficient operations. Remember, safety is everyone's responsibility.

PARKING LOT SAFETY AND GARBAGE MANAGEMENT

Parking lots can be hazardous areas if proper safety measures are not followed. Ensuring the safety of employees and visitors, as well as maintaining cleanliness, is crucial. This talk will cover general parking lot safety, the importance of proper garbage disposal, specific winter safety tips, and maintaining visibility at night.

Key Points

1. **General Parking Lot Safety**:
   - **Speed Limits**: Adhere to posted speed limits and drive slowly, especially in areas with high pedestrian traffic.
   - **Pedestrian Awareness**: Always be on the lookout for pedestrians. Yield the right of way and use designated crosswalks.
   - **Parking Etiquette**: Park within designated spaces and avoid blocking entrances, exits, or fire lanes.
   - **Vehicle Maintenance**: Ensure your vehicle is in good working condition to prevent breakdowns or accidents in the parking lot.
2. **Garbage Disposal**:
   - **Proper Disposal**: Dispose of garbage in designated bins. Do not leave trash on the ground or in parking spaces.
   - **Hazardous Materials**: Follow proper procedures for disposing of hazardous materials. Do not leave items like batteries, chemicals, or sharp objects in the parking lot.

- **Cleanliness**: A clean parking lot reduces the risk of accidents and creates a more pleasant environment for everyone.

3. **Winter Safety**:
   - **Snow and Ice Removal**: Ensure that snow and ice are promptly removed from parking lots and walkways. Use salt or sand to improve traction.
   - **Visibility**: Wear high-visibility clothing and ensure your vehicle's lights are on in low-light conditions.
   - **Footwear**: Wear appropriate footwear with good traction to prevent slips and falls.
   - **Vehicle Preparation**: Keep your vehicle winter-ready with proper tires, antifreeze, and a full tank of gas.
4. **Maintaining Visibility at Night**:
   - **Lighting**: Ensure parking lots are well-lit. Report any broken or malfunctioning lights to maintenance.
   - **Reflective Gear**: Wear reflective clothing or accessories to increase your visibility to drivers.
   - **Vehicle Lights**: Use your vehicle's headlights and ensure they are clean and functioning properly.
   - **Stay Alert**: Be extra cautious and aware of your surroundings when walking or driving in the parking lot at night.
5. **See Something, Say Something**:
   - **Report Hazards**: If you see any hazards, such as icy patches, broken lights, or garbage, report them immediately to your supervisor or maintenance team.
   - **Encourage Vigilance**: Foster a culture of vigilance where everyone feels responsible for maintaining safety and cleanliness.
   - **Take Action**: If you can safely address a minor issue, such as picking up trash or spreading salt on an icy patch, do so.

Group Activity: Parking Lot Safety Walk

To reinforce the importance of parking lot safety, we will conduct a safety walk. This activity will help everyone become familiar with potential hazards and the importance of maintaining a safe and clean parking lot.

Instructions:

1. **Form Teams**: Divide into small teams.
2. **Safety Checklist**: Each team will receive a checklist of items to inspect, such as:
   - Checking for proper lighting and reporting any broken lights.
   - Identifying and reporting any icy patches or areas needing salt/sand.
   - Ensuring garbage is properly disposed of and reporting any litter.
   - Observing and noting any vehicles parked improperly or blocking access.
3. **Time Limit**: Set a time limit for the safety walk (e.g., 10 minutes).
4. **Completion**: Teams will inspect the parking lot using the checklist and report their findings.

5. **Debrief**: After the safety walk, gather to discuss what was found and reinforce the importance of each safety measure.

Conclusion
Parking lot safety and cleanliness are essential for preventing accidents and creating a safe environment for everyone. By following general safety guidelines, properly disposing of garbage, taking extra precautions during winter, and maintaining visibility at night, we can reduce risks and ensure a safer workplace. Remember, if you see something, say something. Your vigilance can make a significant difference.

PEDESTRIAN SAFETY IN THE WAREHOUSE OR ON THE JOB SITE

Pedestrian safety is crucial in a warehouse or job site setting where there is constant movement of people, vehicles, and equipment. Ensuring that everyone is aware of the safety protocols can prevent accidents and injuries.

Key Points

1. **Awareness and Visibility**:
   - Always be aware of your surroundings and stay alert.
   - Wear high-visibility clothing to ensure you are seen by operators of vehicles and machinery.
2. **Designated Walkways**:
   - Use designated pedestrian walkways and crossings.
   - Avoid shortcuts through areas where vehicles and machinery are operating.
3. **Communication**:
   - Make eye contact with vehicle operators before crossing their path.
   - Use hand signals or other agreed-upon signals to communicate your intentions.
4. **Vehicle and Equipment Safety**:
   - Operators should always check blind spots and use mirrors.
   - Vehicles should have functioning lights and alarms to alert pedestrians.
5. **Signage and Barriers**:
   - Pay attention to and obey all safety signs and barriers.
   - Report any missing or damaged signs to your supervisor.

Scavenger Hunt Activity
To reinforce the importance of pedestrian safety, we will conduct a scavenger hunt. This activity will help familiarize everyone with the safety features and protocols in our workplace.

Instructions:

1. **Form Teams**: Divide into small teams.
2. **List of Items**: Each team will receive a list of safety-related items and locations to find, such as:

- Designated pedestrian walkways
- High-visibility clothing storage
- Safety signs (e.g., "Pedestrian Crossing," "No Entry")
- Emergency exits.
- First aid kits
- Vehicle mirrors and blind spot indicators

3. **Time Limit**: Set a time limit for the scavenger hunt (e.g., 20 minutes).
4. **Completion**: Teams must take a photo or note the location of each item on the list.
5. **Debrief**: After the scavenger hunt, gather to discuss what was found and reinforce the importance of each safety feature.

Conclusion
Pedestrian safety is a shared responsibility. By staying alert, using designated walkways, and communicating effectively, we can create a safer work environment for everyone. Let us commit to following these safety practices and looking out for each other.

PEOPLE DEVELOPMENT

People development is a critical responsibility for leaders. Investing in the growth and development of your team not only enhances their skills and performance but also contributes to the overall success of the organization. This toolbox talk will cover key aspects of people development and how leaders can effectively support their team's growth.

Importance of People Development

- **Enhance Skills and Knowledge**: Continuous learning and development help employees improve their skills and stay updated with industry trends.
- **Boosts Morale and Engagement**: Employees who feel supported in their development are more engaged and motivated.
- **Increased Retention**: Investing in employee development reduces turnover by showing that the organization values its people.
- **Drives Organizational Success**: A well-developed team contributes to higher productivity and better business outcomes.

Key Strategies for People Development

1. **Identify Development Needs**:
   - Conduct regular assessments to identify the skills and knowledge gaps within your team. Use performance reviews, feedback sessions, and self-assessments to gather insights.
2. **Create Development Plans**:
   - Develop personalized growth plans for each team member. Set clear, achievable goals and outline the steps needed to reach them. Include opportunities for training, mentorship, and on-the-job learning.

3. **Provide Training and Resources**:
    - Offer access to training programs, workshops, and online courses. Ensure that employees have the resources they need to develop their skills, such as books, software, and tools.
4. **Encourage Continuous Learning**:
    - Foster a culture of continuous learning by encouraging employees to seek out new knowledge and skills. Promote the idea that learning is a lifelong process and support their efforts to grow.
5. **Offer Regular Feedback and Coaching**:
    - Provide constructive feedback and coaching to help employees improve. Regularly check in on their progress and offer guidance and support as needed.
6. **Recognize and Reward Development**:
    - Acknowledge and reward employees' efforts and achievements in their development. Recognition can be a powerful motivator and reinforces the importance of growth.

Conclusion
People development is a vital aspect of leadership that benefits both employees and the organization. By identifying development needs, creating personalized plans, providing training and resources, encouraging continuous learning, offering feedback and coaching, and recognizing achievements, leaders can effectively support their team's growth. Let's commit to fostering a culture of development and helping our team members reach their full potential. Stay supportive and stay engaged!

PERSONAL SAFE 2 HOME

Our Safety Mission can be summarized by the phrase, "Safe 2 Home." But what does that phrase really mean? The simple fact is, we all have other people who count on us to make it home. Whether it is a spouse, children, grandchildren, parents, brothers, sisters, or even Uncle Hank (that we cannot actually stand). The fact, people need you.

Because we are loved and people need us, it is essential that we make getting home each day with all our fingers and toes a priority. It is also equally important that we recognize that those we work with, and those we interact with each day, have their own Safe 2 Home mission. The trick is, how can we help ourselves and others achieve that incredibly unique goal? At the end of the day, our favorite cat, dog, goldfish is waiting for us to feed them.

Discuss:
Go around the group and ask each other to share what their personal Safe 2 Home mission is. This is a fantastic opportunity to connect with each other.

Your Safety is on you:
Now you recognize the importance of making Safety part of your daily routine, let us identify the responsibilities we each have.

1. It is our job to make sure we understand the tasks we are assigned. It is okay to ask for further instructions or clarification.

2. Follow all safety policies, procedures, and warnings. No one wants to work with someone who ignores the safety rules. After all, we all want to make it Safe 2 Home.
3. If you have not been trained or authorized to do a task or use equipment, engage a supervisor and ask for help.
4. See something, say something. Report on all unsafe situations and actions immediately.
5. If an incident occurs, ensure the safety of others in the area and prevent any further injuries before addressing the injured. Do not leave the area, ask someone else to leave and seek help if possible.
6. PPE is not just recommended, it is required. Use it, wear it, and protect yourself.
7. Guards and just as important as using the correct tool for the job. Do not go around, remove, or alter guards or use equipment that has been altered. Report equipment that needs to be repaired or replaced.
8. Know and understand company reporting procedures. First Report of Injury reporting policies are posted at common gathering points. All incidents should be reported according to incident and accident reporting policies.

Discuss:
Where can you find equipment operating procedures, incident reporting procedures, or contact information for our Safety Team leaders?

POWER TOOL SAFETY

Portable electric tools are essential in many workplaces. But they can pose significant hazards if not used properly. This talk will cover key safety practices to ensure safe operation and maintenance of these tools.

General Safety Tips

1. **Inspection Before Use**:
   - Always inspect tools for damage before use. Check for frayed cords, loose parts, and any signs of wear.
   - Ensure that guards are in place and function correctly.
2. **Proper Use**:
   - Use the right tool for the job. Do not use a tool for a purpose it was not designed for.
   - Follow the manufacturer's instructions for operation and maintenance.
3. **Personal Protective Equipment (PPE)**:
   - Wear appropriate PPE such as safety glasses, gloves, and hearing protection.
   - Ensure clothing is not loose to avoid entanglement.
4. **Handling and Storage**:
   - Disconnect tools when not in use, before servicing, and when changing accessories.
   - Store tools in a dry, secure place to prevent damage and unauthorized use.

Discuss:
In what tasks do you find a need for power tools? What is your typical inspection routine and how religious are you about following that process every time? How would you report safety concerns if you identify with power tools? Is there any PPE that is needed?

PERSONAL PROTECTIVE EQUIPMENT

The last line of Safety Prevention is Safety Protection. We call this Personal Protective Equipment or PPE. PPE is specifically created to defeat potential hazards that may exist within the workplace or worksite. As a physical layer of protection from potential hazards, PPE is often sized or fitted not only for the job but also to the individual.

Discuss:
What are the items we would commonly consider PPE? Hard hats, footwear, eye protection, gloves, etc.

For what tasks might you use PPE today?

PPE Care:
Ensure that PPE is inspected before each use and remains in good working condition. It must also be properly stored when not in use. For example, respiratory PPE such as KN95 masks are designed to be a onetime use item. One time use consists of a standard 8-hour day. A mask used longer than 8 hours no longer provides adequate protection. Respiratory aids that include various filter types also have listed hours of use or exposure. Meaning once the package containing the filters is opened, the clock starts for its usable life. Proper storage can increase the useable life of a filter. Like most PPE, please follow care guidelines included with the PPE.

Common Types of PPE:

Gloves: Wearing gloves for jobs/tasks that require additional hand protection is essential. Gloves protect our hands from cuts, punctions, abrasions, and burns. Gloves also can drastically increase your grip. Grips can improve lifting strength and sustainability by as much as 80%. Worn or damaged gloves should be disposed of and replaced.

Eye/Face Protection: Safety glasses, goggles, and shields are worn to protect the eyes and face. Various hazards, including flying debris, chemical splash/gas, dust, and spatter can be mitigated through the use of PPE.

Foot Protection: Closed-toed shoes provide a base level of protection. It's important to utilize footwear designed to assist and increase safety from hazards. Footwear is available to improve slip resistance, protect toes, and ankles.

Hearing Protection: Certain positions also require hearing protection, due to the length of exposure to mild to excessive noises throughout the facility. Various kinds of hearing protection are available. Disposable hearing protection should be disposed of and replaced after each removal from the ear canal. Over the ear devices should be cleaned, stored, and maintained correctly.

PPE is the last line of defense. Are there areas where we rely on PPE today, which could be changed to eliminate the danger? How can I improve the odds of making it Safe 2 Home?

POWERED INDUSTRIAL VEHICLES (TUGGERS)

Cart tuggers are essential for moving heavy loads efficiently in a warehouse setting. However, improper use can lead to accidents and injuries. Understanding and following safety protocols is crucial for the safe operation of cart tuggers.

Key Points

1. **Pre-Operational Checks**:
   - **Inspect the Tugger**: Check for any visible damage, such as worn tires, loose parts, or leaks.
   - **Battery and Fuel**: Ensure the battery is fully charged or the fuel tank is full.
   - **Controls and Brakes**: Test the controls and brakes to ensure they are functioning properly.
2. **Safe Driving Practices**:
   - **Speed Limits**: Adhere to the warehouse speed limits. Drive at a safe speed, especially around corners and in areas with high pedestrian traffic.
   - **Clear Visibility**: Ensure you have a clear view of your path. Use mirrors and look in all directions before moving.
   - **Avoid Distractions**: Stay focused on driving. Avoid using phones or other distractions while operating the tugger.
3. **Load Handling**:
   - **Secure Loads**: Ensure that all loads are securely attached to the tugger. Use straps or other securing devices if necessary.
   - **Even Distribution**: Distribute the weight evenly to prevent tipping or instability.
   - **Height Restrictions**: Do not stack loads too high, as this can obstruct your view and increase the risk of tipping.
4. **Emergency Procedures**:
   - **Stop Safely**: In case of an emergency, stop the tugger safely and turn off the engine.
   - **Report Incidents**: Report any accidents or near-misses to your supervisor immediately.
   - **First Aid**: Know the location of first aid kits and how to use them in case of an injury.
5. **Training and Certification**:
   - **Proper Training**: Ensure all operators are properly trained and certified to use cart tuggers.
   - **Refresher Courses**: Participate in regular refresher courses to stay updated on safety protocols and best practices.

Discussion Points

1. What are the common hazards associated with operating cart tuggers in our warehouse?
2. How can we improve communication between tugger operators and pedestrians?
3. What steps should be taken if a tugger malfunctions during operation?

Conclusion

The safe operation of cart tuggers is essential for preventing accidents and ensuring a smooth workflow in the warehouse. By conducting pre-operational checks, following safe driving practices, handling loads correctly, and prioritizing pedestrian safety, we can minimize risks and maintain a safe working environment. Let's commit to these safety protocols and look out for each other.

PRESSURE TO PERFORM

The pressure to perform can be intense in any workplace. While striving for high performance is important, excessive pressure can lead to negative outcomes, including rushing, which can compromise safety and quality.

Understanding the Pressure to Perform

- **High Expectations**: Meeting deadlines and achieving targets can create significant pressure.
- **Competition**: Competitive environments can push employees to work faster and harder.
- **Fear of Failure**: The fear of not meeting expectations can drive employees to take shortcuts.

Dangers of Rushing Due to Performance Pressure

- **Increased Risk of Accidents**: Rushing to meet deadlines can lead to overlooking safety procedures and result in accidents.
- **Compromised Quality**: Hurrying through tasks often leads to mistakes and lower quality work.
- **Stress and Burnout**: Constant pressure to perform can cause stress and fatigue, impacting overall well-being and productivity.

Discussion Points

1. **Recognizing Performance Pressure**:
   - What are some signs that you or your colleagues are feeling pressured to perform? How can we identify when this pressure is leading to unsafe behaviors, such as rushing?

2. **Impact on Safety and Quality**:
   - How does the pressure to perform affect our ability to follow safety protocols and maintain high-quality work? Can you share an example where performance pressure led to rushing and a subsequent mistake or near-miss?
3. **Strategies to Manage Pressure**:
   - What strategies can we implement to manage performance pressure effectively? How can we support each other in maintaining a balanced approach to work that prioritizes safety and quality over speed?

Addressing Safety Concerns Due to Management Pressure

- **Communicate Openly**: If you feel that management pressure is compromising safety, communicate your concerns clearly and respectfully to your supervisor.
- **Document Concerns**: Keep a record of specific instances where safety protocols were ignored due to pressure. This documentation can be useful when discussing the issue with management.
- **Seek Support**: Talk to your safety officer or human resources department if you feel that your concerns are not being addressed. They can provide additional support and escalate the issue if necessary.
- **Use Safety Committees**: Participate in or bring up your concerns in safety committee meetings where collective voices can have a stronger impact.
- **Know Your Rights**: Familiarize yourself with workplace safety regulations and your rights as an employee. You have the right to a safe working environment.

Conclusion

While the pressure to perform is a common aspect of many workplaces, it's important to manage this pressure to avoid negative outcomes like rushing. By recognizing the signs of performance pressure and discussing strategies to manage it, we can create a safer and more productive work environment. If management pressure is resulting in safety concerns, communicate openly, document issues, seek support, and know your rights. Let's commit to supporting each other in maintaining a balanced approach to work. Stay mindful and stay safe!

PROPANE DANGERS

Propane is a widely used fuel for heating, cooking, and powering equipment. However, it poses significant hazards if not handled properly. This talk will cover the basics of propane, the dangers associated with it, proper storage and connection of propane tanks, and emergency response procedures.

What Is Propane?

Propane is a flammable hydrocarbon gas that is commonly used as fuel. It is stored under pressure in liquid form in tanks and cylinders.

Dangers of Propane

- **Flammability**: Propane is highly flammable and can ignite easily from sparks, open flames, or hot surfaces.
- **Asphyxiation**: In enclosed spaces, propane can displace oxygen, leading to asphyxiation.
- **Explosion Risk**: Leaking propane can accumulate and create an explosive atmosphere.

Storage of Propane Tanks

- **Approved Containers**: Store propane in approved, clearly labeled tanks and cylinders.
- **Ventilation**: Store tanks in a well-ventilated area away from living spaces and ignition sources.
- **Temperature Control**: Keep tanks in a cool, dry place to prevent pressure buildup and leaks.
- **Upright Position**: Store tanks upright to ensure the pressure relief valve functions correctly.
- **Regular Inspection**: Inspect tanks regularly for signs of damage, rust, or leaks.

Connecting Propane Tanks

- **Check Connections**: Ensure all connections are tight and secure before using propane.
- **Use Proper Equipment**: Use regulators and hoses designed for propane use.
- **Leak Detection**: Apply a soap and water solution to connections to check for leaks. Bubbles indicate a leak.
- **Turn Off When Not in Use**: Always turn off the propane supply when the equipment is not in use.

Emergency Response

- **Leak Detection**: If you smell propane (a rotten egg odor), immediately evacuate the area and call emergency services.
- **Fire Response**: In case of a propane fire, use a Class B fire extinguisher if it is safe to do so. Otherwise, evacuate and call emergency services.
- **Medical Response**: If someone is exposed to propane, move them to fresh air immediately. If they have trouble breathing, seek medical attention.

Safety Data Sheet (SDS)
For detailed information on propane safety, please refer to the Safety Data Sheet (SDS) provided by your supplier. [Link to SDS Sheet]

Interactive Questions

1. What are some common hazards associated with propane that you have encountered in your work?
2. How do you ensure that propane tanks are stored properly in your workplace?
3. Can you describe the steps you would take if you detect a propane leak?
4. What precautions do you take when connecting propane tanks to equipment?
5. How do you handle and dispose of propane tanks in your work area?

Conclusion
Understanding the hazards associated with propane and following proper safety practices can prevent accidents and injuries. Always handle propane with care, use appropriate protective equipment, and follow storage and connection guidelines to ensure a safe working environment. Stay vigilant and report any safety concerns to your supervisor.

PROTECTING YOUR SLEEP

Getting proper rest is essential for maintaining overall health, safety, and productivity. Let's discuss the importance of protecting your sleep, including the benefits of good sleep, symptoms of sleep problems, and suggestions for improving sleep quality.

1. **Benefits of Good Sleep**
   - **Improved Concentration and Productivity**: Quality sleep enhances cognitive function, helping you stay focused and efficient at work.
   - **Better Mood**: Adequate rest can improve your mood and reduce irritability, contributing to a more positive work environment.
   - **Enhanced Physical Health**: Good sleep supports immune function, reduces the risk of chronic illnesses, and promotes overall physical well-being.
   - **Increased Safety**: Well-rested individuals are more alert and less likely to make mistakes or have accidents.
   - **Discussion**: How do you feel after a good night's sleep? Can you share any personal benefits you've noticed?
2. **Symptoms of Sleep Problems**
   - **Difficulty Falling Asleep**: Struggling to fall asleep or stay asleep throughout the night.
   - **Daytime Fatigue**: Feeling excessively tired during the day, even after a full night's sleep.
   - **Mood Changes**: Increased irritability, anxiety, or depression.
   - **Decreased Performance**: Difficulty concentrating, memory problems, and reduced productivity.
   - **Discussion**: Have you experienced any of these symptoms? How do they affect your daily life and work?

3. **Suggestions for Improving Sleep Quality**
    - **Establish a Sleep Routine**: Go to bed and wake up at the same time every day, even on weekends.
    - **Create a Relaxing Bedtime Routine**: Engage in calming activities before bed, such as reading, taking a warm bath, or practicing relaxation techniques.
    - **Optimize Your Sleep Environment**: Ensure your bedroom is cool, dark, and quiet. Invest in a comfortable mattress and pillows.
    - **Limit Screen Time**: Avoid screens (phones, tablets, computers) at least an hour before bed, as the blue light can interfere with sleep.
    - **Watch Your Diet**: Avoid large meals, caffeine, and alcohol close to bedtime.
    - **Stay Active**: Regular physical activity can help you fall asleep faster and enjoy deeper sleep.
    - **Manage Stress**: Practice stress-reducing techniques such as deep breathing, meditation, or yoga.
    - **Discussion**: What strategies have you found helpful for improving your sleep? Are there any new tips you'd like to try?

Conclusion

Protecting your sleep is crucial for your health, safety, and productivity. By understanding the benefits of good sleep, recognizing the symptoms of sleep problems, and implementing strategies to improve sleep quality, we can all enjoy better rest and a more productive work environment.

RACKING SAFETY

Racking systems are essential for efficient storage and organization in warehouses and industrial settings. However, improper use or maintenance of racking can lead to serious accidents, including collapses and injuries. Understanding and following racking safety protocols is crucial for maintaining a safe work environment.

Key Points:

1. **Proper Installation**:
    - Ensure racking systems are installed according to the manufacturer's specifications.
    - Use qualified personnel for installation and any modifications.
    - Verify that the racking is level and securely anchored to the floor.
2. **Load Limits**:
    - Adhere to the load limits specified by the manufacturer.
    - Do not overload racks, as this can compromise their structural integrity.
    - Distribute weight evenly across the shelves to prevent tipping or collapse.
3. **Regular Inspections**:
    - Conduct regular inspections of racking systems to identify any damage or wear.

- Look for signs of bending, rust, or loose bolts.
- Report any damage immediately and do not use damaged racks until they are repaired.

4. **Safe Loading and Unloading**:
   - Use appropriate equipment, such as forklifts or pallet jacks, for loading and unloading.
   - Ensure that loads are stable and secure before moving them.
   - Avoid climbing on racks or using them as ladders.
5. **Clear Aisles and Access**:
   - Keep aisles clear of obstructions to allow safe access to racking systems.
   - Ensure there is adequate space for equipment to maneuver safely.
   - Mark pedestrian walkways clearly to separate them from areas where forklifts operate.
6. **Training and Awareness**:
   - Provide training for all employees on racking safety protocols.
   - Ensure that employees understand the importance of following safety guidelines.
   - Encourage a culture of safety where employees feel comfortable reporting hazards.

Conclusion

Racking safety is essential for preventing accidents and ensuring a safe workplace. By following proper installation, load limits, regular inspections, safe loading and unloading practices, and maintaining clear aisles, we can minimize the risk of racking-related incidents. Let's commit to adhering to these safety protocols and looking out for each other.

If you have any questions or need further information, please feel free to ask!

Activity: Racking Safety Inspection

To reinforce the importance of racking safety, we will conduct a hands-on inspection activity. This will help everyone become familiar with identifying potential hazards and understanding the importance of regular inspections.

Instructions:

1. **Form Teams**: Divide into small teams.
2. **Inspection Checklist**: Each team will receive a checklist of items to inspect, such as:
   - Checking for any visible damage to racks (bending, rust, loose bolts).
   - Verifying that load limits are clearly marked and adhered to.
   - Ensuring that aisles are clear of obstructions.
   - Confirming that racks are properly anchored and level.
3. **Conduct the Inspection**: Teams will inspect a designated area of the warehouse or job site using the checklist.
4. **Report Findings**: After the inspection, teams will report their findings to the group. Discuss any issues found and how they can be addressed.

READY TO WORK

Being prepared for work is essential for safety, productivity, and professionalism.

Importance:

- **Safety**: Reduces accident risks.
- **Productivity**: Ensures efficient workflow.
- **Professionalism**: Demonstrates commitment and reliability.

Workplace Expectations:

1. **Punctuality**: Arrive on time.
2. **Proper Attire**: Wear appropriate clothing and PPE.
3. **Readiness**: Be well-rested and focused.
4. **Tools**: Bring necessary tools and check their condition.
5. **Attitude**: Maintain a positive and cooperative attitude.
6. **Task Knowledge**: Know your tasks and safety protocols.

Discussion Points:

1. **Personal Preparation**: What routines help you stay ready for work?
2. **Team Support**: How can we support each other in meeting expectations?
3. **Challenges**: What challenges do you face, and how can we address them?

Conclusion
Showing up ready to work is crucial for a safe and efficient workplace. Let's commit to being prepared and supporting each other. Stay prepared and stay safe!

RECOGNIZING FALL DANGERS

Working at height presents unique risks and requires careful planning and safety measures. Today, we'll have a group-led discussion on best practices for working at height, including working from floors with unprotected edges, ladders, elevated equipment, and roofs. Let's share our experiences and knowledge to ensure everyone's safety.

Discussion Points:

1. **Working from Floors with Unprotected Edges**
   - **Best Practices**:
     - Always use guardrails or temporary barriers to protect edges.
     - Keep the work area clean and free of trip hazards.
     - Use warning signs to alert others of the hazard.
   - **PPE Required**: Safety harnesses, lanyards, and hard hats.
   - **Discussion**: Have you encountered unprotected edges in your work? What measures did you take to stay safe?

2. **Ladders**
    - **Best Practices**:
        - Inspect ladders before use to ensure they are in good condition.
        - Set up ladders on stable, level ground.
        - Maintain three points of contact (two hands and one foot or two feet and one hand) while climbing.
        - Do not overreach or lean too far to the side.
    - **PPE Required**: Non-slip footwear, hard hats.
    - **Discussion**: What are some common mistakes people make when using ladders? How can we avoid them?
3. **Elevated Equipment (e.g., Scaffolding, Aerial Lifts)**
    - **Best Practices**:
        - Ensure equipment is properly assembled and inspected before use.
        - Use guardrails and fall arrest systems.
        - Follow manufacturer guidelines for safe operation.
        - Do not exceed weight limits.
    - **PPE Required**: Safety harnesses, lanyards, hard hats, and sometimes safety glasses.
    - **Discussion**: What challenges have you faced when working with elevated equipment? How did you address them?
4. **Roofs**
    - **Best Practices**:
        - Use fall protection systems such as guardrails, safety nets, or personal fall arrest systems.
        - Be aware of weather conditions that can make roofs slippery.
        - Avoid working near roof edges without proper protection.
        - Ensure roof access points are secure.
    - **PPE Required**: Safety harnesses, lanyards, hard hats, and sometimes safety shoes.
    - **Discussion**: What precautions do you take when working on roofs? Have you ever had a close call, and what did you learn from it?

Conclusion

Working at height requires vigilance and adherence to safety protocols. By sharing our experiences and best practices, we can help each other stay safe. Remember to always use the appropriate PPE and follow safety guidelines to prevent accidents.

RECYCLING

Recycling at work is an important practice that helps reduce waste, conserve resources, and promote environmental sustainability. By recycling properly, we can make a positive impact on our environment and contribute to a greener workplace.

Key Points:

1. **Know What to Recycle**:
   - **Paper**: Office paper, newspapers, magazines, and cardboard.
   - **Plastics**: Bottles, containers, and packaging labeled with recycling symbols.
   - **Metals**: Aluminum cans, tin cans, and other metal items.
   - **Glass**: Bottles and jars.
2. **Proper Disposal**:
   - **Separate Materials**: Use designated bins for different types of recyclables. Ensure items are clean and free of food residue.
   - **Follow Guidelines**: Adhere to local recycling guidelines and company policies for proper disposal.
3. **Reduce and Reuse**:
   - **Reduce Waste**: Minimize the use of disposable items and opt for reusable alternatives.
   - **Reuse Items**: Repurpose materials whenever possible before recycling.
4. **Encourage Participation**:
   - **Educate**: Inform colleagues about the benefits of recycling and how to do it correctly.
   - **Lead by Example**: Demonstrate good recycling habits and encourage others to follow suit.

Conclusion
Recycling at work is a simple yet effective way to contribute to environmental sustainability. By knowing what to recycle, disposing of materials properly, reducing waste, and encouraging participation, we can create a greener and more responsible workplace. Let's commit to recycling and making a positive impact on our environment.

REFLEX REACTION

Ever wonder why a healthcare worker taps on your knee with a rubber mallet? They are trying to inspire a natural reflex reaction. Your leg should kick forward seemingly on its own.

We are born with "pre wired" reactions built into our internal operating systems. Most of these are located in the spinal cord, but some are tied to major motor functions in your brain.

Our Bodies and Brains are designed to keep us alive and protect us from dangers. This natural tendency has been engrained in our bodies over centuries.

The challenges start however, when our bodies work to inspire a reflex reaction and that action endangers us.

Let's discuss an example.

(John is pushing a broom through the prep area. While sweeping Jill pulls into the area on her order picker. As Jill rounds the corner safely away from John, a large box on her picker starts to tip off. John races over to try to secure the load and…)

There are several possible outcomes in this situation. Can you talk about some of the potential hazards John could come into contact with in this scenario?

- Approaching a moving picker?
- How heavy is the object?
- Are there other people or machines present?

This is a very dangerous situation. In most cases, reaction based activity doesn't allow our brains long enough to analyze the risk in a given situation. Let's be on the lookout for Reflex dangers.

As a group, discuss one or two other situations such as the one described above. When might you encounter a reflex reaction that could cause an injury?

Watch out for each other and let's all make it home safe.

RESPIRATORY SAFETY

Respiratory protection is essential in environments where airborne contaminants, such as dust, fumes, gases, and vapors, pose a risk to health (29 CFR 1910.134, Respiratory Protection, 2025). The following Safety Data Sheet (SDS) recommendations ensure that we use the correct respiratory protection for specific hazards.

Key Points:

1. **Understanding Respiratory Hazards**:
   - Airborne contaminants can cause respiratory issues, including irritation, lung damage, and long-term health problems.
   - Common hazards include dust, chemical fumes, vapors, and biological agents.
2. **Types of Respiratory Protection**:
   - **Disposable Masks**: Suitable for protection against dust and non-toxic particles.
   - **Half-Face Respirators**: Provide protection against dust, fumes, and certain gases and vapors when equipped with the appropriate filters.
   - **Full-Face Respirators**: Offer additional protection for the eyes and face, suitable for more hazardous environments.
   - **Powered Air-Purifying Respirators (PAPRs)**: Use a battery-powered blower to provide clean air, ideal for extended use in high-contamination areas.
3. **Following SDS Recommendations**:
   - **Identify Hazards**: Review the SDS for each chemical or material you work with to understand the specific respiratory hazards.
   - **Select Appropriate Protection**: Use the SDS to determine the type of respiratory protection required. The SDS will specify the necessary filter type and protection level.
   - **Fit Testing**: Ensure respirators fit properly to provide effective protection. Conduct regular fit tests as recommended.

- **Maintenance and Care**: Follow the SDS and manufacturer's instructions for cleaning, storing, and maintaining respiratory equipment. Replace filters and cartridges as needed.

4. **Proper Use of Respiratory Protection**:
   - **Donning and Doffing**: Follow proper procedures for putting on and removing respirators to avoid contamination.
   - **Seal Check**: Perform a seal check each time you wear a respirator to ensure it fits correctly and provides a proper seal.
   - **Training**: Participate in training sessions to understand how to use and maintain respiratory protection effectively.

Conclusion

Respiratory protection is vital for safeguarding our health in environments with airborne hazards. By following SDS recommendations and using the appropriate respiratory protection, we can prevent respiratory issues and ensure a safe working environment. Always review the SDS for guidance and adhere to safety protocols.

RUSHING

Rushing through tasks can lead to mistakes, accidents, and injuries. It's important to understand the dangers of rushing and how to maintain a steady, safe pace at work.

Dangers of Rushing:

- **Increased Risk of Accidents**: Rushing can cause you to overlook safety procedures, leading to accidents and injuries.
- **Poor Quality of Work**: Hurrying through tasks often results in mistakes and subpar work quality.
- **Stress and Fatigue**: Constantly rushing can lead to increased stress and fatigue, which further impacts safety and performance.

Discussion Points:

1. **Identifying Situations Where We Rush**:
   - What are some common situations in our workplace where we feel the need to rush? How can we recognize these moments and remind ourselves to slow down?
2. **Impact on Safety and Quality**:
   - How does rushing affect our ability to follow safety protocols and produce high-quality work? Can you share an example of where rushing led to a mistake or near-miss?
3. **Strategies to Avoid Rushing**:
   - What strategies can we implement to manage our time better and avoid the need to rush? How can we support each other in maintaining a steady and safe workplace?

Conclusion

Rushing can have serious consequences for safety, work quality, and overall well-being. By recognizing the dangers and discussing ways to avoid rushing, we can create a safer and more productive work environment. Let's commit to working at a steady pace and supporting each other in this effort. Stay mindful and stay safe!

STACKING AND STORAGE SAFETY

Proper stacking and storage of materials are crucial to maintaining a safe and efficient workplace. Incorrect stacking can lead to accidents, injuries, and damage to materials.

Discussion Points:

1. **Personal Protective Equipment (PPE)**
   - **Wear gloves** to protect your hands from sharp edges and splinters.
   - **Safety glasses** to shield your eyes from dust and debris.
2. **General Stacking Guidelines**
   - **Stack materials in designated areas** to avoid blocking walkways and exits.
   - **Ensure stable and level surfaces** for stacking to prevent tipping.
   - **Do not stack materials too high**; keep stacks at a manageable height to avoid collapse.
3. **Specific Material Handling**
   - **Palletized goods:** : Stack no more than two pallets high and ensure the upper Pallet is squarely on the lower Pallet .
   - **Timber**: Use dunnage to keep stacks level. Store different lengths separately.
   - **Pipes**: Stack small pipes in racks and secure large diameter pipes at the base. Avoid pyramid stacking.
   - **Prefabricated Panels**: Stack flat or in suitable racks. Do not lean.
4. **Using Pallets**
   - **Inspect pallets** for damage before use.
   - **Secure loads properly** on pallets to prevent shifting.
   - **Use the correct size and type** of pallet for the load.
5. **Manual Handling**
   - **Assess the load** before lifting to determine if mechanical assistance is needed.
   - **Use proper lifting techniques** to avoid strain and injury.
   - **Work in teams** for heavy or awkward loads.
6. **Storage Safety**
   - **Keep storage areas clean and organized** to prevent tripping hazards.
   - **Label storage areas** clearly to ensure materials are returned to the correct place.
   - **Regularly inspect** storage areas for potential hazards.

Conclusion
By following these guidelines, we can reduce the risk of accidents and injuries related to stacking and storage. Remember, safety is a shared responsibility, and proper practices help keep everyone safe.

SAFETY RESPONSIBILITIES

We all have a responsibility to demonstrate genuine concern for the health and well-being of those we work with. Generally, most of our interactions with others happen at work. It's no surprise that prevention is key to keeping each other safe. Ultimately, the employer is responsible for the health and safety of their staff and guests, however, it takes everyone to work together to keep people safe.

Safety by the Numbers
Only 20% of incidents that occur are due to equipment failure. In other words, 80% of incidents are due to human errors in process execution or behavior. Meaning nearly all incidents can be prevented with general care, concern, and training.

Secret Sauce
Believe it or not, there isn't one. All we must do to prevent injuries is follow established procedures and watch out for each other. That's it. Always refer to the Standard Operating Procedure. If the way a task is to be performed could be made safer, review proposed changes with your supervisor. Never change the SOP until the changes recommended are approved.

Not only must the process be followed, but the design of tools and equipment is also done intentionally. Modifying or changing tools or equipment to perform tasks differently or different tasks entirely should be left to the engineers.

Be a Safety Leader
Don't just follow the leader, be one. If you see something, say something. Your suggestions and recommendations matter. A natural tendency we all share is that we become averse to change and blind to every day. Sometimes it takes an unexpected recommendation or reminder from someone else to make real change. Your opinion matters.

Safety is everyone's responsibility.

SANITIZE TO PREVENT ILLNESS

Welcome, everyone! Today, we're going to talk about something that's crucial for our health and productivity. We keep our workspaces and hands clean to prevent the spread of germs. With cold and flu season upon us, it's more important than ever to maintain good hygiene practices. Let's dive into some key points to help us stay healthy and safe.

Key Points:

1. **Importance of Sanitizing Workspaces**:
   - Germs can survive on surfaces for hours or even days. Regularly disinfecting high-touch areas like desks, keyboards, and doorknobs can significantly reduce the spread of germs. Use EPA-approved disinfectants and follow the instructions for effective cleaning.
2. **Hand Hygiene**:
   - Washing your hands with soap and water for at least 20 seconds is one of the best ways to prevent the spread of germs. If soap and water aren't available, use a hand sanitizer with at least 60% alcohol. Remember to wash your hands before eating, after using the restroom, and after coughing or sneezing.
3. **Staying Home When Sick**:
   - If you're feeling unwell, it's important to stay home to avoid spreading illness to your coworkers. This helps protect everyone and ensures a healthier work environment. Notify your supervisor and follow company policies regarding sick leave.
4. **Promoting a Clean Environment**:
   - Encourage everyone to keep their workspaces tidy and free of clutter. Provide sanitizing wipes and hand sanitizers at convenient locations around the office. Regularly restock these supplies to ensure they are always available.

Discussion Points:

- **What are some effective ways we can remind each other to sanitize our workspaces regularly?**
- **How can we support our colleagues in staying home when they are sick without feeling pressured to come to work?**

SCISSOR LIFT SAFETY

Scissor lifts are versatile tools used for working at heights, but they come with inherent risks. Proper maintenance and safe operation are crucial to prevent accidents and ensure the safety of operators and those around them.

Discussion Prompt for Scissor Lift Training:

Topic: Enhancing Safety and Efficiency in Scissor Lift Operations

Prompt: "Let's discuss the key aspects of scissor lift safety and maintenance. Think about your experiences and observations while operating or maintaining scissor lifts. What are some best practices you've found effective in ensuring safety? Have you encountered any challenges or near-misses that highlighted the importance of certain safety protocols? How do you ensure that pre-operational checks and maintenance are performed consistently and thoroughly? Let's share our insights and strategies to improve our overall safety and efficiency when using scissor lifts."

Feel free to share your thoughts and experience.

Here are some additional items to help improve your next experience working with or on a scissor lift.

Key Points:

1. **Pre-Operational Checks**:
   - **Visual Inspection**: Check for any visible damage, such as cracks, leaks, or worn parts.
   - **Controls and Safety Devices**: Test all controls and safety devices to ensure they are functioning correctly.
   - **Fluid Levels**: Check hydraulic fluid levels and look for any leaks.
   - **Tires and Wheels**: Inspect tires and wheels for wear and proper inflation.
2. **Safe Operation**:
   - **Training**: Only trained and authorized personnel should operate scissor lifts.
   - **Load Limits**: Adhere to the manufacturer's load limits. Do not exceed the maximum weight capacity.
   - **Stability**: Ensure the lift is on a stable, level surface before raising the platform. Use outriggers if available.
   - **Guardrails**: Always use guardrails and ensure they are in place before operating the lift.
3. **Maintenance Safety**:
   - LOTO **Procedures**: Follow LOTO procedures to ensure the lift is not accidentally operated during maintenance.
   - **Manufacturer's Guidelines**: Perform maintenance according to the manufacturer's guidelines and schedule.
   - **Replacement Parts**: Use only manufacturer-approved replacement parts to maintain the integrity of the lift.
   - **Documentation**: Keep detailed records of all maintenance and inspections.
4. **Emergency Procedures**:
   - **Emergency Lowering**: Know how to use the emergency lowering mechanism in case of a power failure.
   - **First Aid**: Be familiar with the location of first aid kits and emergency contact numbers.
   - **Incident Reporting**: Report any accidents or near-misses to your supervisor immediately.
5. **Environmental Awareness**:
   - **Weather Conditions**: Avoid using scissor lifts in high winds, rain, or other adverse weather conditions.
   - **Overhead Hazards**: Be aware of overhead hazards such as power lines and low ceilings.
   - **Surrounding Area**: Ensure the area around the lift is clear of obstacles and personnel.

Conclusion

Maintaining and operating scissor lifts safely is essential to prevent accidents and injuries. By conducting thorough pre-operational checks, following safe operation protocols, adhering to maintenance guidelines, and being aware of your environment, we can ensure a safe working environment for everyone. Let's commit to these safety practices and look out for each other.

SEASONAL CHANGES

Hello everyone! As the seasons change, so do the safety challenges we face at work. Today, we'll discuss essential safety tips for each season to help us stay safe and prepared all year round. Let's dive into some key points for each season and talk about how sudden weather changes can impact on our safety.

Winter Safety:

1. **Preventing Slips and Falls**:
   - Keep walkways clear of snow and ice. Use salt or sand to improve traction and wear appropriate footwear with good grip.
2. **Heating Safety**:
   - Ensure all heating equipment is properly maintained and used safely. Keep space heaters away from flammable materials and never leave them unattended.

Spring Safety:

1. **Allergy Awareness**:
   - Be mindful of increased pollen and allergens. Ensure proper ventilation and consider using air purifiers to maintain good indoor air quality.
2. **Wet Conditions**:
   - Spring showers can create slippery conditions. Use caution when walking on wet surfaces and keeping entryways dry by using mats and signage.

Summer Safety:

1. **Heat Stress Prevention**:
   - Stay hydrated and take regular breaks in shaded or cool areas. Wear light-weight, breathable clothing and use sunscreen to protect against UV rays.
2. **Outdoor Safety**:
   - Be aware of insects and pests. Use insect repellent and check for ticks after spending time outdoors.

Autumn Safety:

1. **Leaf Hazards**:
   - Wet leaves can be as slippery as ice. Keep walkways clear of leaves and use caution when walking on leaf-covered surfaces.

2. **Shorter Days**:
    - With less daylight, ensure adequate lighting in work areas and be extra cautious when driving or walking in low-light conditions.

Discussion Point:

- **Sudden Weather Changes**:
    - Weather can change rapidly, especially during transitional seasons. Sudden storms, temperature drops, or unexpected snow can create hazardous conditions. Always check the weather forecast and be prepared for sudden changes. Discuss how we can stay informed and ready to adapt to these changes to ensure our safety.

SEIZURES

Seizures can occur suddenly and unexpectedly. Recognizing the symptoms and knowing how to respond can help ensure the safety of the affected individual. Individuals who regularly experience Seizures should inform their supervisor so proper medical responses can be planned.

Symptoms:

- **Uncontrollable Jerking Movements**: Sudden, repetitive movements of the arms and legs.
- **Staring Spells**: A period of unresponsiveness or staring into space.
- **Loss of Consciousness**: The person may collapse or become unresponsive.
- **Confusion**: Disorientation or confusion after the seizure.
- **Sudden Falls**: Falling without any apparent reason.

Response:

1. **Stay Calm**: Keep yourself and others calm.
2. **Protect the Person**: Move objects away to prevent injury. Cushion their head if possible.
3. **Time of the Seizure**: Note the duration of the seizure.
4. **Do Not Restrain**: Do not hold the person down or try to stop their movements.
5. **Do Not Put Anything in Their Mouth**: This can cause injury.
6. **Turn Them on Their Side**: If possible, turn the person on their side to help keep their airway clear.
7. **Call 911**: If the seizure lasts more than 5 minutes, the person has multiple seizures, or if they are injured during the seizure.

Conclusion

Recognizing the symptoms of a seizure and responding appropriately can help ensure the safety of the affected individual. Stay informed and be ready to act in an emergency. Stay safe!

SEXUAL HARASSMENT

Sexual harassment in the workplace is a serious issue that affects everyone. It is important to understand what it is, how to prevent it, and what to do if it happens. This talk will cover key points about sexual harassment and include open discussion points to ensure everyone understands their role in maintaining a respectful and safe work environment.

What Is Sexual Harassment?
Sexual harassment includes any unwelcome sexual advances, requests for sexual favors, and other verbal or physical conduct of a sexual nature that creates a hostile or offensive work environment. **Discussion Point**: Can anyone provide examples of behaviors that might be considered sexual harassment?

Types of Sexual Harassment:

1. **Quid Pro Quo**: This happens when someone asks for sexual favors in exchange for job benefits.
2. **Hostile Work Environment**: This happens when unwelcome sexual behavior makes the workplace intimidating, hostile, or offensive.
3. **Discussion Point**: How can we recognize the difference between a quid pro quo and a hostile work environment?

Preventing Sexual Harassment:

- **Respect and Professionalism**: Treat all coworkers with respect and always maintain professionalism.
- **Clear Communication**: Communicate boundaries clearly and respect the boundaries set by others.
- **Education and Training**: Participate in regular training sessions on sexual harassment prevention.
- **Discussion Point**: What steps can we take as a team to foster a respectable and inclusive workplace?

Reporting Sexual Harassment:

- **Know the Policy**: Familiarize yourself with the company's sexual harassment policy and reporting procedures.
- **Speak Up**: If you experience or witness sexual harassment, report it immediately to your supervisor or HR department.
- **Support Each Other**: Encourage and support coworkers who come forward with complaints.
- **Discussion Point**: What are the barriers that might prevent someone from reporting sexual harassment, and how can we address them?

Responding to Sexual Harassment:

- **Listen and Support**: If a coworker confides in you about experiencing harassment, listen without judgment and offer support.
- **Document Incidents**: Keep a record of any incidents, including dates, times, and descriptions of what occurred.
- **Follow Up**: Ensure that reported incidents are followed up on and that appropriate actions are taken.
- **Discussion Point**: How can we create a supportive environment that encourages reporting and addresses harassment effectively?

Conclusion

Sexual harassment has no place in our workplace. By understanding what it is, taking steps to prevent it, and knowing how to respond, we can create a safe and respectful environment for everyone. Let us commit to fostering a culture of respect and support.

SHOULDERS, KNEES, AND ELBOWS

Ergonomics is hugely important in what we do in every facet of the organization. It plays perhaps the most significant role for Prep and Delivery Teams. Today, let's focus on how our Shoulders, Knees, and Elbows are directly affected by poor ergonomics and discuss some things we can do to protect ourselves.

Shoulders:

- Posture: Maintain a neutral posture. Keep your shoulders relaxed and avoid hunching.
- Lifting: Use your legs, not your back or shoulders. Keep the load close to your body.
- Repetitive movements: Take breaks to avoid repetitive strain, stretch your shoulders regularly.

Knees:

- Lifting: Bend your knees, not your back, when lifting. This reduces strain on your lower back and knees. Remember to "hinge" at your joints.
- Standing: If you stand for long periods, shift your weight between your legs and use anti-fatigue mats if available.
- Kneeling: Use knee pads or a cushion to protect your knees, a piece of Styrofoam sheeting works in a pinch.

Elbows:

- Workstation Setup: Ensure your workstation is set up so your elbows are at a 90-degree angle when typing or using tools.
- Repetitive Tasks: Avoid repetitive elbow movements. Take breaks and stretch your arms.
- Lifting: Keep your elbows close to your body when lifting to reduce strain.

General Tips:

- Stretching: Incorporate regular stretching into your routine to keep your muscles flexible.
- Breaks: Take frequent breaks to rest and rest your posture.
- Ergonomic Equipment: Use ergonomic tools like lift tables, rolling carts, and skates to reduce body strain.

Conclusion

By paying attention to our shoulders, knees, and elbows, we can significantly reduce the risk of injury and improve our comfort at work. Remember, proper ergonomics is not just about preventing injuries but also about enhancing our overall productivity and wellbeing.

SITE SECURITY

Hello everyone! Today, we're going to discuss the importance of security in our store, warehouse, and site areas. Ensuring a secure environment is crucial for protecting our assets, maintaining safety, and preventing unauthorized access. Let's go over some key points to keep in mind.

Key Points:

1. **Access Control**:
   - Ensure that only authorized personnel have access to secure areas. Use ID badges, key cards, or other access control measures to monitor and restrict entry.
2. **Physical Security Measures**:
   - Implement physical barriers such as fencing, locks, and secure storage containers. Regularly inspect these measures to ensure they are in good condition and functioning properly.
3. **Surveillance and Monitoring**:
   - We Utilize Surveillance Systems to monitor activity in and around the premises. We ensure that these systems are regularly maintained, and that footage is reviewed as needed.
4. **See Something, Say Something**:
   - Encourage all employees to be vigilant and report any suspicious activity or security concerns immediately. If you see something unusual or out of place, say something to your supervisor or the designated security officer.

Employee Activity:

Security Walkthrough:

- Pair up with a colleague and conduct a security walkthrough of your area. Look for potential security vulnerabilities such as unlocked doors, obstructed surveillance cameras, or areas that lack proper lighting.
- Take notes on any issues you find and discuss them with your partner. After the walkthrough, report your findings to your supervisor or the security officer.
- This activity helps everyone become more aware of security measures and encourages proactive identification and resolution of potential security risks.

Discussion Points:

- **How can we improve our current security measures to better protect our site?**
- **What are some examples of suspicious activities that should be reported immediately?**

SLIP, TRIP, FALL PREVENTION

Good morning, everyone! I hope you are all doing well today. I wanted to take a few minutes to talk about something that affects all of us, whether we are working in retail, an office, or a warehouse: preventing slips, trips, and falls. These incidents can happen to anyone, and they often result in injuries that could have been avoided. By being aware and taking some simple precautions, we can keep ourselves, our colleagues, and our customers safe. Let us dive into some practical tips for different environments.

Retail Customer Safety:

1. **Clean and Dry Floors**: Ensure floors are clean and dry. Immediately clean up spills and place wet floor signs to alert customers.
2. **Clear Aisles**: Keep aisles free of clutter and obstacles. Ensure merchandise is properly shelved and not protruding into walkways.
3. **Proper Lighting**: Ensure all areas, especially stairways and entrances, are well lit to help customers see where they are walking.
4. **Safe Entryways**: Use mats at entrances to absorb moisture and prevent slippery surfaces during wet weather.

Retail/Office Professionals:

1. **Organized Workspaces**: Keep workspaces tidy. Store items properly and avoid leaving boxes or equipment in walkways.
2. **Cords and Wires**: Secure cords and wires away from walkways. Use cable covers or tape them down to prevent tripping hazards.

3. **Proper Footwear**: Wear appropriate footwear with non-slip soles, especially in areas prone to spills or wet floors.
4. **Report Hazards**: Encourage employees to report any hazards immediately, such as loose carpeting or uneven flooring.

General Warehouse:

1. **Housekeeping**: Maintain a clean and organized warehouse. Regularly inspect and clean floors to remove debris and spills.
2. **Clear Pathways**: Ensure pathways are clear of obstructions. Mark designated walkways and keep them free of pallets and equipment.
3. **Proper Storage**: Store materials securely and at appropriate heights to prevent items from falling and creating tripping hazards.
4. **Use of Equipment**: Use ladders and step stools safely. Ensure they are in good condition and used correctly to prevent falls.

Conclusion
Eliminating slips, trips, and falls demands dedication and consistent focus. By maintaining clean and organized environments, using proper equipment, and encouraging a culture of safety, we can significantly reduce the risk of these incidents. Let us all commit to making our workplaces safer for everyone.

SNOW SHOVELING

Shoveling snow is a common task during the winter months, but it can be physically demanding and potentially hazardous. It is important to follow safe practices to prevent injuries and ensure the job is done efficiently.

Safety Tips for Shoveling Snow:

- **Warm Up**: Before you start, warm up your muscles with light stretching or a short walk to prevent strains.
- **Dress Appropriately**: Wear layered clothing to stay warm and choose waterproof boots with good traction to prevent slips.
- **Use the Right Shovel**: Select a lightweight, ergonomic shovel to reduce strain on your back.
- **Lift Properly**: Bend your knees, keep your back straight, and lift with your legs. Avoid twisting your body while lifting.
- **Push do not Lift**: Whenever possible, push the snow to the side rather than lifting it. This reduces the strain on your body.
- **Take Breaks**: Shoveling can be exhausting. Take regular breaks to rest and hydrate.
- **Stay Hydrated**: Drink plenty of water to stay hydrated, even in chilly weather.

Knowing When to Use Larger Equipment or Ask for Help:

- **Heavy Snowfall**: If the snowfall is heavy or the area to be cleared is large, consider using a snow blower or other larger equipment to make the job easier and safer.
- **Physical Limitations**: If you have any health conditions or physical limitations, do not hesitate to ask for help from a neighbor, friend, or professional service.
- **Time Constraints**: If you are short of time, using larger equipment or getting help can ensure the job is done quickly and efficiently without compromising safety.

Conclusion
Shoveling snow can be a strenuous task, but by following these safety tips and knowing when to use larger equipment or ask for help, you can prevent injuries and complete the job safely. Stay warm, stay safe, and take care of yourself during the winter months!

SPEAKING WITH RESPECT

Effective communication is essential in any workplace. Speaking with respect fosters a positive work environment, enhances teamwork, and improves overall productivity. This talk will cover the importance of respectful communication and provide tips for maintaining professionalism in our interactions.

Key Points:

1. **Understanding Respectful Communication**:
    - Respectful communication involves listening actively, speaking kindly, and valuing others' opinions.
    - It means avoiding negative language, sarcasm, and dismissive attitudes.
2. **Benefits of Speaking with Respect**:
    - **Improved Relationships**: Respectful communication builds trust and strengthens relationships among colleagues.
    - **Enhanced Teamwork**: When team members feel respected, they are more likely to collaborate effectively and support each other.
    - **Positive Work Environment**: A respectful atmosphere reduces conflicts and creates a more enjoyable workplace for everyone.
3. **Tips for Professional Dialog**:
    - **Listen Actively**: Pay attention to what others are saying without interruption. Show that you value their input.
    - **Use Positive Language**: Choose words that are encouraging and constructive. Avoid negative or harsh language.
    - **Be Mindful of Tone**: Your tone of voice can convey respect or disrespect. Aim for a calm and friendly tone.

- **Give Constructive Feedback**: When providing feedback, focus on the behavior or issue, not the person. Offer solutions and support.
- **Acknowledge Differences**: Respect diverse perspectives and backgrounds. Be open to different viewpoints and ideas.

4. **Handling Difficult Conversations**:
   - **Stay Calm**: Keep your emotions in check and approach the conversation with a clear mind.
   - **Be Clear and Direct**: Communicate your points clearly and respectfully. Avoid beating around the bush.
   - **Seek Solutions**: Focus on finding a resolution rather than placing blame.

Questions to Discuss:

1. What are some examples of respectful communication you have experienced in our workplace?
2. How can we improve our communication to ensure it is always respectful?
3. What challenges do you face when trying to communicate respectfully, and how can we address them?
4. How can we support each other in maintaining a respectful work environment?
5. What steps can we take to handle difficult conversations more effectively?

Conclusion

Speaking with respect is crucial for creating a positive and productive workplace. By practicing active listening, using positive language, and being mindful of our tone, we can foster a respectful and collaborative environment. Let's commit to improving our communication skills and supporting each other in maintaining professionalism.

STEP LADDER SAFETY

Step ladders are commonly used in many workplaces, but improper use can lead to serious injuries. This talk will cover essential safety tips for using step ladders, including maintaining three points of contact and proper storage when not in use.

Safety Precautions:

1. **Inspect Before Use**
   - **Check for Damage**: Inspect the ladder for cracks, bent rungs, or other damage.
   - **Stability**: Ensure the ladder is stable and all locking mechanisms are secure before climbing.
2. **Maintain Three Points of Contact**
   - **Climbing Safely**: Always maintain three points of contact (two hands and one foot, or two feet and one hand) when climbing or descending the ladder.
   - **Avoid Overreaching**: Keep your body centered between the side rails. Do not overreach or lean to one side.

3. **Use the Ladder Properly**
   - **Correct Angle**: Set up the ladder at the correct angle. The base should be one foot away from the wall for every four feet of ladder height.
   - **Top Step**: Do not stand on the top step or the top cap of the ladder.
   - **Load Capacity**: Do not exceed the ladder's maximum load capacity.
4. **Proper Storage**
   - **Clean and Dry**: Clean the ladder and ensure it is dry before storing it.
   - **Secure Storage**: Store the ladder in a secure, dry place where it won't be damaged. Hang it horizontally on a wall or store it vertically in a designated area, using straps or chains to secure it from tipping over.
   - **Prevent Damage**: Avoid storing the ladder in high-traffic areas where it could be knocked over or damaged.

Guided Demonstration:

1. **Inspecting the Ladder**
   - **Visual Check**: Demonstrate how to visually inspect the ladder for any signs of damage, such as cracks or bent rungs.
   - **Stability Check**: Show how to check that all locking mechanisms are secure, and the ladder is stable.
2. **Climbing with Three Points of Contact**
   - **Proper Technique**: Demonstrate climbing the ladder while maintaining three points of contact. Emphasize the importance of keeping two hands and one foot, or two feet and one hand, in contact with the ladder at all times.
   - **Avoiding Overreach**: Show how to keep your body centered between the side rails and avoid leaning to one side.
3. **Setting Up the Ladder**
   - **Correct Angle**: Demonstrate setting up the ladder at the correct angle, with the base one foot away from the wall for every four feet of ladder height.
   - **Top Step Safety**: Show the correct way to use the ladder without standing on the top step or top cap.
4. **Storing the Ladder**
   - **Cleaning and Drying**: Demonstrate how to clean and dry the ladder before storage.
   - **Secure Storage**: Show how to store the ladder securely, either horizontally on a wall or vertically in a designated area, ensuring it is not in a high-traffic area. Use straps or chains to secure it from tipping over or falling.

Conclusion

By following these safety precautions and practicing proper techniques, we can minimize the risks associated with using step ladders and ensure a safer working environment. Always stay vigilant and report any safety concerns to your supervisor.

STRETCH WRAP

Stretch wrap is used in warehouses and shipping environments to protect and secure products. While it is an effective tool for stabilizing loads, improper use can lead to product damage. One key issue is that stretch wrap can tighten over time, potentially damaging goods if not applied correctly.

Key Points:

1. **Risks of Improper Use**:
   - **Over-Tightening**: Applying stretch wrap too tightly can cause it to contract further over time, leading to crushed or damaged products.
   - **Inadequate Wrapping**: Not using enough layers or not securing the wrap properly can result in unstable loads that may shift or collapse.
2. **Proper Application Techniques**:
   - **Consistent Tension**: Apply the stretch wrap with consistent tension. Avoid pulling too tightly, especially around corners and edges.
   - **Multiple Layers**: Use multiple layers of wrap to ensure stability without overtightening. Typically, 3-5 layers are sufficient for most loads.
   - **Secure the Base**: Start wrapping at the base of the pallet and work your way up, ensuring the base is securely wrapped to prevent shifting.
   - **Overlap Layers**: Overlap each layer by about 50% to ensure full coverage and added strength.
3. **Monitoring and Adjusting**:
   - **Regular Checks**: Periodically check wrapped pallets for signs of overtightening or shifting. Adjust the wrap if necessary.
   - **Temperature Changes**: Be aware that temperature changes can affect the tension of the stretch wrap. Monitor wrapped pallets in different environments.
4. **Training and Awareness**:
   - **Employee Training**: Provide training on the proper use of stretch wrap, including the risks of over-tightening and how to apply it correctly.
   - **Best Practices**: Share best practices and tips for using stretch wrap effectively to prevent product damage.

Conclusion

Stretch wrap is a valuable tool for securing products, but it must be used correctly to avoid damaging goods. By applying the wrap with consistent tension, using multiple layers, and regularly monitoring wrapped pallets, we can ensure the safety and integrity of our products. Let's commit to following these best practices and maintaining the quality of our goods.

STRESS

Stress is a common experience that can affect anyone, especially in demanding work environments. Today, we'll discuss what stress is, its effects, how to mitigate it, and how to recognize it in ourselves and others.

Definition of Stress

Stress is the body's response to any demand or challenge. It can be triggered by numerous factors, including work pressures, personal issues, or unexpected changes. While some stress can be motivating, excessive stress can be harmful.

Effects of Stress:

- **Physical Effects**: Headaches, muscle tension, fatigue, sleep disturbances, and weakened immune system.
- **Emotional Effects**: Anxiety, irritability, depression, and mood swings.
- **Cognitive Effects**: Difficulty concentrating, memory problems, and indecisiveness.
- **Behavioral Effects**: Changes in appetite, increased use of alcohol or drugs, and withdrawal from social activities.

How to Mitigate Stress:

- **Healthy Lifestyle**: Maintain a balanced diet, exercise regularly, and get enough sleep.
- **Time Management**: Prioritize tasks, break them into manageable steps, and avoid overcommitting.
- **Relaxation Techniques**: Practice deep breathing, meditation, or yoga to relax your mind and body.
- **Social Support**: Talk to friends, family, or colleagues about your stressors. Sometimes, sharing your concerns can help alleviate stress.
- **Professional Help**: Seek support from a counselor or therapist if stress becomes overwhelming.

Recognizing Stress in Others:

- **Physical Signs**: Noticeable fatigue, frequent headaches, or changes in appearance.
- **Emotional Signs**: Increased irritability, mood swings, or signs of anxiety and depression.
- **Cognitive Signs**: Difficulty concentrating, forgetfulness, or indecisiveness.
- **Behavioral Signs**: Changes in eating or sleeping habits, withdrawal from social interactions, or increased use of substances.

Discussion:

- **Personal Experiences**: Have you experienced stress at work? How did you manage it?
- **Support Strategies**: What can we do as a team to support each other in managing stress?
- **Resources**: Are there any resources or programs available through our workplace to help with stress management?

Conclusion

Understanding and managing stress is crucial for maintaining a healthy and productive work environment. By recognizing the signs of stress and implementing strategies to mitigate it, we can support ourselves and our colleagues in achieving better well-being.

STRETCH, FLEX, AND REST

The alarm's going off- Again. You're up, dressed, and out the door. You walk into work, clock in, and jump right into your shift. The boss is already making rounds, pushing for more production.

Then it happens.

You push a heavy box out of the way—nothing crazy—and the next thing you know, you're in the ER with your arm in a sling and a torn bicep.

So, what went wrong?

You didn't warm up. Your body wasn't ready. And to be honest, this isn't rare. A very large segment of workers' comp injuries come from soft tissue incidents caused by strains and sprains. Lifting, twisting, just moving wrong can cause them. But we can greatly reduce our risk. Let's break it down into three simple tasks.

We need to understand why *stretching matters*, how to *incorporate exercise* into our daily routine, and how to inject *short breaks* in our day to prevent future incidents.

Stretch: Get Set, Go

Stretching is for anyone who moves, yes, that includes you and I. Here's the basics:

Warm up first. Get blood flowing before you start. Walk and swing your arms to get started.

No Bouncing. That quick jerky stretch. Yea, that actually tears the muscle fibers and increases risk of injury. Instead, slowly hold a comfortable stretch for 15-30 seconds. Then release.

Quality over Quantity. You don't need to do 20 different stretches. Just do a few but do them right.

Flex: Strength and Readiness

Flex isn't about showing off, it's about staying in shape, just enough to handle the job. Here are two simple exercises you can do daily to work those major muscle groups.

- Wall Sits: Yep, sitting down counts as exercise. Great for legs and core. Just lean against a wall like you're sitting in a chair. Hold for 30–45 seconds. Stand and repeat several times.
- Arm Circles: Spin those shoulders and loosen that upper back. Do 12–15 spins forward, then backwards. Try rotating both arms at once, in opposite directions just for fun.

These fun exercises will keep you in the game.

Rest: Fatigue Is the Enemy of Injury Prevention

Take 5-minute breaks. Especially when doing repetitive tasks.
Drink water. Help hydrate those muscles. Tight muscles tear easily.
Listen to your body. If something feels off, it probably is.
A short break now can save you weeks of recovery later.

Group Activity: Stretch and Share:

Let's Stretch (30 seconds each)

- Neck Rolls
- Shoulder Stretches
- Calf Raises
- High Knees
- Hand Flexers

Quick Chat:

What part of your body feels tight or sore most often?
What is one thing you can do today to take better care of your body?

STRUCK BY

"Struck by" hazards are one of the leading causes of injuries and fatalities in the workplace. These incidents occur when a worker is hit by an object, equipment, or vehicle. Understanding and mitigating these hazards is crucial for maintaining a safe work environment.

Common "Struck By" Hazards:

- **Falling Objects**: Tools, materials, or debris falling from heights.
- **Moving Equipment**: Forklifts, cranes, and other machinery moving through the work area.
- **Vehicles**: Trucks, cars, and other vehicles operating in or near the work zone.
- **Flying Objects**: Particles or debris ejected from tools or machinery.

Prevention Strategies:

1. **Personal Protective Equipment (PPE)**: Wear hard hats, safety glasses, and high-visibility clothing to protect against falling and flying objects.
2. **Secure Tools and Materials**: Ensure tools and materials are properly secured when working at heights to prevent them from falling.
3. **Establish Safe Zones**: Designate and mark safe zones where workers can stand clear of moving equipment and vehicles.
4. **Use Spotters**: Employ spotters to guide equipment operators and alert workers to potential hazards.

5. **Maintain Equipment**: Regularly inspect and maintain tools and machinery to ensure they are in good working condition and do not pose a hazard.
6. **Training and Awareness**: Provide training on recognizing and avoiding "struck by" hazards. Encourage workers to stay alert and be aware of their surroundings.

Emergency Response:

- **First Aid**: Administer first aid for minor injuries and seek medical attention for more serious injuries.
- **Report Incidents**: Report all "struck by" incidents to supervisors immediately, regardless of severity.
- **Review and Improve**: After an incident, review what happened and implement measures to prevent future occurrences.

Conclusion
"Struck by" hazards are a significant risk in many workplaces, but with proper precautions and awareness, we can reduce the likelihood of these incidents. By wearing appropriate PPE, securing tools and materials, establishing safe zones, using spotters, maintaining equipment, and providing training, we can create a safer work environment. Stay alert and stay safe!

SUMO STANCE LIFTING?

Sumo stance is a term we use to describe the physical positioning we should use when physically lifting products. Feet slightly more than shoulder with apart, knee's bent, shoulders straight, back straight, neck in line with the spine, chin up and head supported. We positioned our body as close to the product as possible, secure the load with a good grip, and pull the load into your body, then lift.

Approaching each lift as if it's the first time, allows us to go through the mental checklist of the Sumo Stance position. Each time ensure our body mechanics are in the proper position to take on the load. Be ready to lift and move, allowing your feet to turn with your body rather than twisting at the waistline. Proper lifting technique puts you in an "Attack" position. Allowing you to *control* the load.

Control is key.

A lack of load control is a leading cause of lifting or carrying injury. When you load a truck, or a trailer securement of the load is essential. Why would lifting a box or carrying a load be any different? Proper control is achieved through using the right equipment and most importantly using the right mindset when approaching the load. Control the Load don't let it control you.

Tips:

- Its estimated that more than 50% of load control, grip, and carry capacity is dependent on hand strength. Give yourself an advantage by wearing gloves with a tacky palm and finger surface. This greatly increases your capacity and reduces the likelihood of injury.

- Approach each lift with natural body mechanics in mind. If you have to move something at floor level, avoid working on your knees. This naturally puts the load stress on the next moving joint, your waist and lower back. Instead kneel and use your planted foot "leg" to assist in moving the load and support your lifting arm your thigh. Taking additional pressure off your lower back.
- Avoid awkward twist maneuvers while lifting or carrying a load. Go around corners, rather than making sudden directional changes. Turn your feet first and rotate your body smoothly in the direction of travel.
- Use mechanical advantage. We have carts of varying types and sizes to provide you with various ways to move the product.
- Avoid carrying products longer than is needed. Use carts for transport of goods when possible.

Conclusion: Remember most strains and sprains are caused by sudden movements, the next leading cause is muscle fatigue. Be properly hydrated, use proper technique and approach each lift as if it is the first time. Alert your supervisor of any potential strain/sprain to prevent larger injuries.

TASK PLANNING

Presenter's Example:
"Let me share a simple example of how I used task planning to organize a quick and efficient team meeting.

Task: Organizing a Team Meeting

1. **Setting Clear Objectives**
   - **Objective**: To discuss the progress of our current project and address any issues.
   - **SMART Goal**: Hold a 30-minute meeting to review project milestones and assign the next steps.
2. **Developing a Detailed Plan**
   - **Steps**:
     1. Choose a date and time that works for all team members.
     2. Book a meeting room or set up a virtual meeting link.
     3. Prepare an agenda outlining key discussion points.
     4. Send out invitations with the agenda attached.
   - **Responsibilities**:
     5. I took charge of scheduling and preparing the agenda.
     6. Team members were asked to review the agenda and come prepared with updates.
3. **Communicating the Plan and Following Up**
   - **Communication**:
     1. Sent out the meeting invitation and agenda via email.
     2. Reminded team members the day before the meeting.

- **Follow-Up**:
  3. During the meeting, I ensured everyone had a chance to speak.
  4. After the meeting, I sent out a summary of the discussion and assigned tasks.

4. **Monitoring and Evaluating Progress**
   - **Monitoring**:
     1. Keep track of the assigned tasks and deadlines.
     2. Held brief check-ins to ensure tasks were on track.
   - **Evaluation**:
     3. Review the outcomes of the meeting and make notes for improvement in future meetings.

By following these steps, I was able to organize a productive meeting that kept everyone informed and on track. This approach can be applied to any task, big or small, to ensure efficient and effective execution.

Conclusion

Using task planning for even simple tasks like organizing a meeting can significantly improve efficiency and outcomes. Applying the same techniques to more complex projects and jobs has the same effect. Let's apply these principles to our daily tasks to enhance our productivity and achieve our goals.

THE ART OF USING A TWO WHEEL DOLLY

Did you know there is a proper way to use a Two Wheel Dolly? This simple yet effective tool can make a big difference in our daily tasks, especially when it comes to moving bulky or heavy items.

Benefits of Using the Two Wheel Dolly:

- Reduces physical strain:
  - A two-wheel dolly helps distribute the weight of heavy objects, reducing strain on your back, shoulders, and arms.
  - This can prevent injuries and combat fatigue, allowing you to work more efficiently and comfortably.
- Increase Efficiency:
  - Using a dolly can speed up the process of moving items, as it allows you to transport multiple items at once.
  - This can save time and effort, making your workflow more productive.
- Improve Safety:
  - By reducing the need to lift heavy objects manually, a dolly minimizes the risk of lifting related injuries.
  - It also provides better control when moving items, reducing the likelihood of dropping or damaging them.

- Versatility:
  - Two-wheel dollies are versatile and can be used for a variety of tasks, from moving boxes and furniture to transporting equipment and supplies.
  - They are easy to maneuver in tight spaces, making them ideal for use in different environments.

How to Use a Two Wheel Dolly:

- Inspect the Dolly:
  - Before using it, check the dolly for any signs of damage or wear. Ensure the wheels are in good condition and the frame is sturdy. Pay particular attention to the lift plate.
  - Make sure the dolly is appropriate for the weight and size of the items you need to move.
- Load the Dolly Properly:
  - Place the items on the dolly with the heaviest items at the bottom to maintain stability.
  - Boxed goods can be tilted forward and the dolly inserted underneath. Using leverage rather than strength to capture the load.
- Use proper technique.
  - When lifting the dolly, bend your knees and keep your back straight to avoid injury.
  - Put on foot on the axle of the dolly and brace your other foot slightly behind you.
  - Push with the foot on the dolly and pull the load towards you until the weight is on the cart and the load is elevated for transport.
  - Keep the dolly and load close. Extending your arms puts an additional load on your arms, shoulders, neck, and back.
- Move with caution:
  - Push the dolly forward, keeping an eye on your path to avoid obstacles or uneven surfaces.
  - Take your time and avoid sudden movements.
  - Maintain balance and control.

Conclusion
Using a Two Wheel Dolly can greatly enhance our ability to move heavy and bulky items safely. By understanding the benefits and following proper usage techniques we can reduce risk and injury. Let's make sure we use this valuable tool to take advantage of the benefits.

THE LITTLE THINGS

Engaging Activity: Start with a quick icebreaker. Ask everyone to share a small task they did recently and how it impacted their day. This will help illustrate the significance of small actions.

"Good morning, everyone! Today, we are going to talk about the importance of doing the little things and doing them right. To kick things off, let us go around and share a small task you did recently and how it made a difference in your day. It could be something as simple as organizing your desk or double-checking a report. Let us hear your stories!"

1. **Attention to Detail**
   - **Why it matters**: Small details can have a big impact on the overall quality and safety of a task.
   - **How to do it**: Double-check your work, follow procedures precisely, and use checklists to ensure nothing is overlooked.
   - **Benefits**: Reduces errors, enhances quality, and ensures safety.
2. **Consistency in Execution**
   - **Why it matters**: Consistent performance builds reliability and trust within the team.
   - **How to do it**: Develop good habits, follow standard operating procedures, and maintain a high standard of work.
   - **Benefits**: Ensures uniformity in output, improves efficiency, and fosters a culture of excellence.
3. **Preventing Small Issues from Becoming Big Problems**
   - **Why it matters**: Addressing minor issues promptly can prevent them from escalating into major problems.
   - **How to do it**: Stay vigilant, report any irregularities immediately, and take corrective actions without delay.
   - **Benefits**: Minimizes downtime, reduces costs, and maintains a safe working environment.
4. **Building a Strong Foundation**
   - **Why it matters**: The success of larger tasks often depends on the proper execution of smaller, foundational tasks.
   - **How to do it**: Focus on getting the basics right, ensure all preliminary steps are completed accurately, and build from a solid base.
   - **Benefits**: Enhances overall project success, improves structural integrity, and boosts team confidence.

Closing Activity

Interactive Exercise: End with a practical exercise. Provide a simple task for everyone to complete, such as organizing a toolbox or creating a checklist for a routine job. Then, discuss how paying attention to the little things made the task easier and more efficient.

"To wrap up, let's put what we've discussed into practice. I will give you a simple task to complete, like organizing a toolbox or creating a checklist for a routine job. Take a few minutes to do this, paying close attention to the details. Once you are done, we will discuss how focusing on the little things made the task easier and more efficient. Let us get started!"

Conclusion: Doing the little things and doing them right is crucial for achieving excellence in our work. By paying attention to detail, maintaining consistency, addressing small issues promptly, and building a strong foundation, we can ensure high-quality outcomes and a safe working environment. Let us commit to these practices and strive for excellence in everything we do. Thank you for your participation!

THE POWER OF THE TBT FOR SUPERVISORS

Toolbox Talks are not just about safety and procedures; they are also a valuable opportunity for supervisors to connect with their team members. By using these talks effectively, supervisors can build trust, improve communication, and foster a positive work environment. This talk will cover how supervisors can use Toolbox Talks to connect with their team.

Importance of Connection:
Building a strong connection with your team helps create a supportive and collaborative work environment. It encourages open communication, boosts morale, and enhances overall productivity.

Discussion Point: How does a strong connection between supervisors and team members benefit our workplace?

Tips for Effective Toolbox Talks:

1. **Be Engaging**: Make the talks interactive by asking questions and encouraging participation. Use real-life examples and stories to illustrate points.
2. **Show Genuine Interest**: Take the time to listen to your team members' concerns and feedback. Show that you value their input and are willing to address their issues.
3. **Be Consistent**: Hold regular Toolbox Talks to maintain ongoing communication and keep safety and other important topics top of mind.
   Discussion Point: What makes a Toolbox Talk engaging and effective for you?

Building Trust:

- **Transparency**: Be open and honest during Toolbox Talks. Share relevant information and updates with your team.
- **Follow Through**: If issues or concerns are raised, ensure they are addressed promptly. Follow up on any commitments made during the talks.
- **Respect**: Treat all team members with respect and acknowledge their contributions.
  Discussion Point: How can supervisors build trust with their team during Toolbox Talks?

Encouraging Open Communication:

- **Create a Safe Space**: Encourage team members to speak up without fear of judgment or retaliation.
- **Active Listening**: Practice active listening by giving your full attention, nodding, and providing feedback.
- **Feedback Loop**: Provide constructive feedback and encourage team members to do the same.
  Discussion Point: What can supervisors do to encourage open communication during Toolbox Talks?

Recognizing and Addressing Concerns:

- **Identify Issues**: Use Toolbox Talks to identify and address any concerns or challenges faced by the team.
- **Problem-Solving**: Collaborate with your team to find solutions to problems. Involve them in the decision-making process.
- **Support**: Offer support and resources to help team members overcome obstacles.
  Discussion Point: How can supervisors effectively recognize and address concerns during Toolbox Talks?

Conclusion
Toolbox Talks are a powerful tool for supervisors to connect with their team members. By being engaging, showing genuine interest, building trust, encouraging open communication, and addressing concerns, supervisors can create a positive and productive work environment. Let us commit to using Toolbox Talks not only for safety but also for building stronger connections within our team.

THINK THROUGH IT

It might seem obvious, but there are some real benefits to thinking through a task before performing it, especially when it comes to lifting and moving products. Many strains and sprains can be prevented simply by thinking through a task before completing it. A clear thought process can help us identify risks, improve efficiency, consider the impact on others, and ensure safety over everyone involved.

Key Points to Consider:

1. Identify risks:
   a. Assess the task and understand the task at hand. What are the steps involved? What tools and equipment will you need, or could you use?
   b. Spot potential hazards associated with the task. This could include physical hazards (Sharp objects, pinch points, or heavy lifting), environmental hazards (slippery surfaces, poor lighting, etc.) and procedural hazards. (Complex machinery, high pressure systems, etc.).

   c. Evaluate the risks or likelihood of injury and the potential severity. Prioritize addressing the highest and most likely risk first.
2. Boost Efficiency:
   a. Plan out the steps necessary to complete the task. Think about the sequence of actions to minimize unnecessary movements or delays.
   b. Gather resources ahead of time to ensure you have all necessary tools, materials, and information before starting. This will prevent interruptions and save time.
   c. Optimize workflows and combine any steps or simplify them if safe to do so. Are there any tools or techniques that make the process faster?
3. Consider the Impact on Others:
   a. Communicate with your team about what you're doing and if it will cause any disruption in their tasks. Communication helps everyone understand and stay prepared.
   b. Coordinate work with others to ensure your task doesn't interfere with theirs. Consider how actions might affect their safety and productivity.
   c. Support: Be ready to assist others if needed. A collaborative approach can improve overall efficiency and safety.
4. Ensure Safety:
   a. Follow all safety protocols and guidelines. Use PPE as required.
   b. Check equipment before starting work. Inspect hand tools, power tools, carts, PPE, and larger equipment before using.
   c. Monitor equipment and your surroundings. Keep work areas clean and organized to prevent incidents.

Conclusion: Establishing your own set process for thinking through an activity before performing it is crucial to identify risks and boost efficiency. By taking the time to plan and communicate, we can reduce the likelihood of incidents and improve the overall function of our departments and company. Let's make it a habit to think before we act and prioritize safety in everything we do.

TIME MANAGEMENT

Maintaining a consistent and efficient work schedule is essential for productivity and workplace harmony. Today, we will discuss the importance of working time, including punching in and out, break guidelines, and minimizing distractions.

Employee-Led Discussion Points:

1. **Punching In and Out**
   - **Why It Is Important**: Accurate timekeeping ensures everyone is paid correctly and helps track work hours for project management.
   - **Best Practices**:
     - Always punch in at the start of your shift and punch out at the end.
     - If you forget, notify your supervisor immediately to correct it.
     - Use the designated time clock or system consistently.

2. **Lunch Breaks and 15-Minute Breaks**
    - **Guidelines**:
        - Take your lunch break at the scheduled time to ensure coverage and fairness.
        - Utilize your 15-minute breaks to rest and recharge.
        - Return to work promptly to maintain workflow and respect your colleagues' time.
    - **Discussion**: How do you feel about the current break schedule? Any suggestions for improvement?
3. **Cell Phone Use and Other Distractions**
    - **Minimizing Distractions**:
        - Limit personal cell phone use to breaks and emergencies.
        - Keep personal calls and texts to a minimum during work hours.
        - Avoid other distractions like social media or non-work-related websites.
    - **Discussion**: What strategies do you use to stay focused? How can we support each other in minimizing distractions?
4. **Consistency Is Key**
    - **Why Consistency Matters**:
        - Consistent practices lead to a more predictable and efficient work environment.
        - Helps in building trust and reliability among team members.
    - **Discussion**: How can we ensure everyone follows these guidelines consistently? What challenges do you face in maintaining consistency?

Conclusion

By adhering to these guidelines and maintaining consistency, we can create a more productive and harmonious workplace. Let us work together to support each other in following these practices.

TOOLBOX TRAINING ACKNOWLEDGEMENT

Training Topic	**Date/Time Presented**
Presented by (Print)	**Location#**
Witnessed by (Print)	**Team/Department**
Attendance to this presentation constitutes training in the above area. By attending you agree you understand that your job duties will be done in accordance with the training. You also understand that a copy of this acknowledgement is kept as part of the permanent file.	
Attendance	
Print	Sign
Print	Sign
Print	Sign

Print	Sign
Print	Sign
Print	Sign
Print	Sign
Print	Sign
Print	Sign
Print	Sign
Print	Sign
Print	Sign
Print	Sign
Print	Sign
Print	Sign
Print	Sign
Print	Sign
Print	Sign
Print	Sign

Thank you for attending this Toolbox Talk. Stay Safe and make it Safe 2 Home.

TORNADO/SEVERE WEATHER SAFETY

Severe weather, including tornadoes, can strike with little warning, posing significant risks to safety. It's crucial to be prepared and know how to respond to protect yourself and others. Following the guidelines provided by NOAA can help us do just that (Tornado Safety Guidelines, 2025).

Key Points:

1. **Understanding Tornadoes and Severe Weather**
    - **Tornadoes**: Violently rotating columns of air extending from a thunderstorm to the ground. They can cause extensive damage and are often accompanied by hail, lightning, and heavy rain.
    - **Severe Weather**: Includes thunderstorms, high winds, hail, and flash floods. Each poses unique hazards and requires specific safety measures.
2. **Preparation and Planning**
    - **Emergency Kit**: Assemble a kit with essentials such as water, non-perishable food, flashlight, batteries, first aid supplies, and important documents.
    - **Communication Plan**: Establish a plan to communicate with family and coworkers. Ensure everyone knows where to go and how to stay in touch during an emergency.

3. **During a Tornado**
    - **Seek Shelter**: Move to a small, windowless interior room on the lowest level of a sturdy building. Avoid mobile homes and vehicles.
    - **Protect Yourself**: Use heavy furniture or a mattress to shield yourself from debris. Cover your head and neck.
4. **During Severe Weather**
    - **Stay Informed**: Monitor weather updates through reliable sources like NOAA Weather Radio, local news, or weather apps.
    - **Avoid Hazardous Areas**: Stay away from windows, doors, and outside walls. Avoid using electrical appliances and plumbing during thunderstorms.
5. **After the Storm**
    - **Assess Safety**: Check for injuries and provide first aid if necessary. Avoid downed power lines and damaged structures.
    - **Report Hazards**: Notify authorities of any hazards such as gas leaks, broken power lines, or structural damage.

Small Group Discussion Questions:

1. **Personal Preparedness**: What items do you have in your emergency kit at home or work? Are there any additional items you think should be included?
2. **Emergency Plans**: How can we improve our current emergency communication and shelter plans at our workplace or home?

Feel free to share your thoughts and experiences. Being prepared can make a significant difference in ensuring everyone's safety during severe weather events.

TWIST AND SHOUT

Good morning, everyone! Today, we're going to talk about a crucial topic that can help prevent injuries and keep us all safe on the job: **Twist and Shout**. We'll cover the dangers of twisting while lifting, how to avoid it, proper lifting techniques, and how to gently remind each other to stay safe.

Part 1: Twist—The Dangers of Twisting While Lifting
Twisting your body while lifting can lead to serious injuries, including:

- **Strained muscles**: Twisting puts extra stress on your muscles, increasing the risk of strains.
- **Herniated discs**: The twisting motion can cause the discs in your spine to slip or rupture.
- **Lower back pain**: Twisting while lifting can lead to chronic lower back pain.

How to Avoid Twisting While Lifting:

1. **Plan your lift**: Before lifting, think about where you're going to place the object and how you'll get there.

2. **Position your body**: Face the object directly and keep your feet shoulder-width apart.
3. **Use your legs**: Bend your knees and use your leg muscles to lift, keeping your back straight.
4. **Move your feet**: Instead of twisting your torso, move your feet to turn your whole body.
5. **Keep the load close**: Hold the object close to your body to maintain balance and reduce strain.

Proper Lifting Techniques:

- **Get a good grip**: Ensure you have a firm hold onto the object before lifting.
- **Lift smoothly**: Avoid jerky movements; lift the object in a smooth, controlled motion.
- **Keep your back straight**: Maintain a neutral spine position throughout the lift.
- **Don't lift too much**: Know your limits and ask for help if the load is too heavy.

Part 2: Shout—How to Gently Remind Others
It's important to look out for each other and ensure everyone is lifting safely. Here are some tips for gently reminding your coworkers:

1. **Be respectful**: Approach your coworker in a friendly and respectful manner.
2. **Use "I" statements**: Instead of saying "You're doing it wrong," try "I noticed you might be twisting while lifting."
3. **Offer help**: Suggest a better way to lift or offer to help with the load.
4. **Share information**: Remind them of the proper lifting techniques we've discussed.
5. **Be supportive**: Encourage a culture of safety where everyone feels comfortable looking out for each other.

Quiz Time!
Let's test our knowledge with a quick quiz:

1. **What are the risks of twisting while lifting?**
   - a) Strained muscles
   - b) Herniated discs
   - c) Lower back pain
   - d) All of the above
2. **What should you do before lifting an object?**
   - a) Twist your body
   - b) Plan your lift
   - c) Lift with your back
   - d) Ignore the weight

3. **How should you position your body when lifting?**
    - a) Face the object directly
    - b) Keep your feet together
    - c) Twist your torso
    - d) Look away from the object
4. **What is a good way to remind a coworker to lift safely?**
    - a) Yell at them
    - b) Use "I" statements
    - c) Ignore them
    - d) Criticize them
5. **Which of the following is a proper lifting technique?**
    - a) Lift with jerky movements
    - b) Keep the load far from your body
    - c) Use your leg muscles
    - d) Twist your back

Conclusion

Remember, safety is everyone's responsibility. By avoiding twisting while lifting and using proper techniques, we can prevent injuries and create a safer workplace. And don't forget to "shout" out to your coworkers in a supportive way to keep everyone safe. Let's stay strong and safe together!

WIND CHILL

Windchill is a critical factor to consider during cold weather as it significantly impacts how cold it feels outside. Understanding windchill and its effects can help prevent cold-related injuries.

What Is Windchill?

- **Definition**: Windchill is the perceived decrease in air temperature felt by the body due to the flow of air. It combines the actual air temperature with the wind speed to determine how cold it feels.
- **Calculation**: Windchill is calculated using a formula that considers both the air temperature and the wind speed. This provides a more accurate measure of the cold's impact on the human body (Windchill Calculator, 2025).

Why Windchill Is More Accurate Than Air Temperature

- **Heat Loss**: Wind increases the rate at which heat is lost from the body. The faster the wind, the quicker the heat loss, making it feel colder than the actual air temperature.
- **Frostbite Risk**: Windchill can accelerate the onset of frostbite. For example, an air temperature of 0°F with a 15-mph wind can feel like -19°F, increasing the risk of frostbite on exposed skin within 30 minutes.

- **Hypothermia**: Prolonged exposure to cold and wind can lead to hypothermia, a dangerous drop in body temperature. Windchill provides a better indication of the risk level compared to air temperature alone.

Safety Tips:

1. **Dress Appropriately**: Wear layers of clothing, including a windproof outer layer. Cover exposed skin to reduce the risk of frostbite.
2. **Limit Exposure**: Minimize time spent outdoors during extreme windchill conditions. Take frequent breaks in a warm shelter.
3. **Stay Informed**: Monitor weather forecasts and windchill warnings. Plan activities accordingly to avoid the coldest and windiest times.

Small Group Discussion:

1. **Workplace Preparedness**: What steps can we take as a team to ensure everyone is protected from windchill during outdoor work?

By understanding and respecting the dangers of windchill, we can better protect ourselves and our coworkers from cold-related injuries. Stay warm and stay safe!

WORKPLACE BULLYING

Workplace bullying is a serious issue that can affect the well-being, productivity, and morale of employees. It's important to recognize, address, and prevent bullying to create a safe and respectful work environment.

Entry Activity: Scenario Discussion:

Activity: "What Would You Do?"

- Present a hypothetical scenario involving workplace bullying.
- Example: "Imagine a coworker is consistently making sarcastic comments about another team member's work in meetings. How would you handle this situation?"
- Discuss as a group how to address the situation effectively.
- Purpose: This activity encourages participants to think critically about handling bullying and promotes discussion on appropriate responses.

Key Points:

1. **Understanding Workplace Bullying**
   - **Definition**: Repeated, health-harming mistreatment of one or more people by one or more perpetrators. It includes verbal abuse, offensive conduct/behaviors, and work interference.
   - **Forms of Bullying**: Includes spreading rumors, undermining work, exclusion, and intimidation.

2. **Impact of Bullying**
   - **On Individuals**: Causes stress, anxiety, depression, and decreased job satisfaction.
   - **On the Workplace**: Leads to reduced productivity, increased absenteeism, and higher turnover rates.
3. **Recognizing Bullying**
   - **Signs**: Unexplained absences, decreased performance, and changes in behavior.
   - **Examples**: Public humiliation, constant criticism, and isolation from team activities.
4. **Preventing and Addressing Bullying**
   - **Policies and Procedures**: Implement clear anti-bullying policies and reporting mechanisms.
   - **Support Systems**: Provide access to counseling and support for affected employees.
   - **Training**: Conduct regular training sessions on recognizing and preventing bullying.
5. **Creating a Respectful Workplace**
   - **Promote Inclusivity**: Encourage teamwork and respect for diversity.
   - **Lead by Example**: Managers and leaders should model respectful behavior.
   - **Open Communication**: Foster an environment where employees feel safe to speak up.

Exit Activity: Reflection and Commitment:

Activity: "Commitment Pledge"

- Each participant writes down one action they will take to help prevent workplace bullying.
- Participants share their pledges with the group.
- Purpose: This activity reinforces the importance of individual responsibility and commitment to creating a respectable workplace.

Small Group Discussion Questions:

1. **Personal Experience**: Have you witnessed or experienced workplace bullying? How was it handled, and what could have been done differently?
2. **Improvement Ideas**: What steps can we take as a team to ensure a bullying-free workplace?

By understanding the impact of workplace bullying and committing to respectful behavior, we can create a positive and supportive work environment for everyone.

WORKPLACE VIOLENCE

Workplace violence is a serious issue that can affect the safety and well-being of employees. It's essential to recognize, prevent, and address any form of violence to maintain a safe and productive work environment.

Entry Activity: Scenario Discussion:

Activity: "What Would You Do?"

- Present a hypothetical scenario involving workplace violence.
- Example: "Imagine a coworker becomes verbally aggressive during a disagreement. How would you handle this situation?"
- Discuss as a group how to address the situation effectively.
- Purpose: This activity encourages participants to think critically about handling violence and promotes discussion on appropriate responses.

Key Points:

1. **Understanding Workplace Violence**
   - **Definition**: Any act or threat of physical violence, harassment, intimidation, or other threatening disruptive behavior that occurs at the work site.
   - **Forms of Violence**: Includes physical assaults, verbal abuse, threats, and harassment.
2. **Impact of Workplace Violence**
   - **On Individuals**: Causes physical injury, emotional trauma, stress, and decreased job satisfaction.
   - **On the Workplace**: Leads to reduced productivity, increased absenteeism, higher turnover rates, and a negative work environment.
3. **Recognizing Workplace Violence**
   - **Signs**: Unexplained injuries, changes in behavior, frequent conflicts, and expressions of anger or frustration.
   - **Examples**: Physical altercations, verbal threats, and intimidating behavior.
4. **Preventing and Addressing Workplace Violence**
   - **Policies and Procedures**: Implement clear anti-violence policies and reporting mechanisms.
   - **Training**: Conduct regular training sessions on recognizing and preventing workplace violence.
   - **Support Systems**: Provide access to counseling and support for affected employees.
   - **Emergency Plans**: Develop and practice emergency response plans for violent incidents.
5. **Creating a Safe Workplace**
   - **Promote Respect**: Encourage a culture of respect and open communication.
   - **Lead by Example**: Managers and leaders should model non-violent behavior.
   - **Report Concerns**: Encourage employees to report any concerns or incidents of violence immediately.

Exit Activity: Reflection and Commitment:

Activity: "Commitment Pledge"

- Each participant writes down one action they will take to help prevent workplace violence.

- Participants share their pledges with the group.
- Purpose: This activity reinforces the importance of individual responsibility and commitment to creating a safe workplace.

Small Group Discussion Questions:

1. **Personal Experience**: Have you witnessed or experienced workplace violence? How was it handled, and what could have been done differently?
2. **Improvement Ideas**: What steps can we take as a team to ensure a violence-free workplace?

By understanding the impact of workplace violence and committing to respectful behavior, we can create a positive and supportive work environment for everyone.

Works Cited

29 *CFR 1910.120 Hazardous Waste Operations and Emergency Response*. (2025). Retrieved from Occupational Safety and Health Administration: www.osha.gov

29 *CFR 1910.132 PPE*. (2025). Retrieved from Occupational Safety and Health Administration: www.osha.gov

29 *CFR 1910.134, Respiratory Protection*. (2025). Retrieved from Occupational Safety and Health Administration: www.osha.gov

29 *CFR 1910.147, The Control of Hazardous Energy (Lockout/Tagout)*. (2025). Retrieved from Occupational Safety and Health Administration: www.osha.gov

29 *CFR 1910.212 General Requirements of all Machines*. (2025). Retrieved from Occupational Safety and Health Administration: www.osha.gov

29 *CFR 1910.333*. (2025). Retrieved from Occupational and Health Administration: www.osha.gov

29 *CFR 1926.453*. (2025). Retrieved from Occupational Safety and Health Administration: www.osha.gov

29 *CFR 1910.1030*. (2025). Retrieved from Occupational Safety and Health Administration: www.osha.gov

29 *CFR 1910.1200 Hazard Communication Standard*. (2025). Retrieved from Occupational Safety and Health Administration: www.osha.gov

AED implementation Guide. . (2020). Retrieved from American Heart Association: www.heart.org

Association, A. B. (2024). *ameriburn.org/resources/burn-incidence-fact-sheet/*. Retrieved February 12, 2025

Battery Recycling and Disposal Guidelines. (2025). Retrieved from Environmental Protection Agency.: www.epa.gov

Cold Stress Guide. (2025). Retrieved from Occupational Safety and Health Administration: www.osha.gov

Federal Size Regulations for Commercial Motor Vehicles. (2025). Retrieved from Federal Highway Administration: www.fhwa.dot.gov

General Safety Requirements for All Machines 29.CFR 1910.212. (2025). Retrieved from Occupational Safety and Health Administration: www.osha.gov

Hand and Portable Powered Tools 29 CFR 1910.242. (2025). Retrieved from Occupational Safety and Health Administration: www.osha.gov

Maxwell, J. C. (2010). *Everyone Communicates, Few Connect: What the Most Effective People Do Differently*. Harper Collins.

NFPA 70E: Standard for Electrical Safety in the Workplace. (2021). Retrieved from National Fire Protection Association: www.nfpa.org

Occupational Safety and Health Administration, OSHA Guidelines for Job Hazard Analysis. (2025, July 31). Retrieved July 31, 2025

Step on Safety. (2022). *Tool Box Talk 32*. 32-Safe-Stacking-of-Materials.pdf

Tornado Safety Guidelines. (2025). Retrieved from National Oceanic and Atmospheric Administration.: www.noaa.gov

Vinson, Daniel C., Vineesha Arelli. (2006). State Anger and the Risk of Injury: A Case Control and Case Crossover Study. *Annals of Family Medicine, Vol. 4, no. 1*, 63–68.

Windchill Calculator. (2025). Retrieved from National Weather Service: www.weather.gov

Index